BASIC ESD AND
I/O DESIGN

BASIC ESD AND I/O DESIGN

SANJAY DABRAL
TIMOTHY MALONEY
Intel Corporation
Santa Clara, California

A Wiley Interscience Publication
JOHN WILEY & SONS, INC.
New York / Chichester / Weinheim / Brisbane / Singapore / Toronto

Library of Congress Cataloging-in-Publication Data:

Dabral, Sanjay.
    Basic ESD and I/O design / Sanjay Dabral and Timothy Maloney.
        p.   cm.
    "A Wiley-Interscience publication."
    Includes bibliographical references and index.
    ISBN 0-471-25359-6 (alk. paper)
    1. Integrated circuits—Very large scale integration—protection.
2. Electronic circuit design.   3. Electric discharges.   4. Static
eliminators.   I. Maloney, Timothy J.
    TK7874.75.D33   1998
    621.39′5—dc21   97-32241

CONTENTS

PREFACE

This effort started as an answer to the numerous questions the authors have repeatedly had to answer about electrostatic discharge (ESD) protection and input/output (I/O) designs. In the past no comprehensive book existed sufficiently covering these areas, and these topics were rarely taught in engineering schools. Thus first-time I/O and ESD protection designers have had considerable trouble getting started. This book is in part an answer to such needs.

As mentioned, I/O and ESD are barely addressed in standard electrical engineering curricula today. However, there is a growing need for high-speed I/O designs, and it is only a matter of time before it becomes imperative to provide students with some exposure to this type of design in the schools. This book is intended to provide a ready reference for such very large scale integrated (VLSI) design and fabrication courses. Also, students using VLSI foundry services for their courses and research will find this book useful in understanding I/O circuits and their testing. In addition, this book can be a resource to anyone with an interest in the subject, even if just a general interest.

This book will also be of use to those whose primary charter is not I/O and ESD design but core VLSI design or computer architecture (i.e., academicians or those in industry). It should help them to understand the key parameters in I/O and ESD designs, enabling them to communicate effectively and make good design and architecture trade-offs.

It should be noted that this is a rapidly growing and dynamic field. Any specific designs described today will probably be outdated by the time of printing. Therefore, general and basic concepts are emphasized in preference to particular solutions. Designability has been stressed, so it is hoped that these principles will still be valid for some time to come.

People designing devices for future technologies will find the material relevant. The impact of ESD and reliability concerns on I/O choices is repeatedly illustrated, especially in the areas of capacitive loading, voltage compatibility, and hot-electron reliability. It is hoped that this text will help process designers better understand the requirements of I/O designers. This should benefit mutual design concerns.

A deliberate attempt has been made to integrate material from various sources, especially material that is only available in scattered text. Subjects that are well covered in other standard texts are referenced and described only so that continuity of thought is preserved, but for the most part they have not been repeated. The authors believe that the references will do better justice to those topics.

This book is intended to be an integrated I/O and ESD text. However, for readers inclined towards I/O design, Chapters 1 and 4–7 will be more meaningful. Readers seeking information on ESD should refer to Chapters 1, 2, 3, 7, and 8.

SANJAY DABRAL
TIMOTHY J. MALONEY

Santa Clara, CA
July, 1998

ACKNOWLEDGMENTS

We gratefully acknowledge the help of Ruth Flores in editing this book. She spent considerable effort and time in revising, suggesting, and finding references that greatly aided in making the text clearer and consistent.

We would also thank Krishna Seshan, Gautam Verma, and Anil Pant for reviewing the book and providing suggestions that have helped simplify the text. We are also grateful to Dilip Sampath, Ming Zeng, and Randy Mooney for reviewing the I/O-related chapters.

S.D.
T. J. M.

CHAPTER 1

INTRODUCTION

1.1. HISTORICAL PERSPECTIVE

Today in industry, electrostatic discharge (ESD) is primarily looked upon as a reliability threat. High-voltage ESD such as lightning damages buildings, outdoor equipment, and power lines, whereas smaller ESDs can play havoc with poorly protected semiconductor devices and electronics. In such an environment, one quickly loses the overall perspective on ESD and wishes that it did not exist. However, *electrostatic discharge* has been a key ingredient in activities such as the following:

- the discovery of electricity, as shown by Benjamin Franklin's kite experiment [Gale, 1994];
- the discovery of wireless communication, in which Hertz used an ESD as a transmitter and another ESD gap as the receiver [Lee, 1996];
- the perpetuation of life by fixing nitrogen in the atmosphere to nitrates, which helps plants to grow [McGraw-Hill, 1982]; and
- a cleaner atmosphere, by helping electrostatically clean the exhaust gases produced by factories [van Nostrand, 1995]

In view of some of these benefits in the past and those to come in the future, the problems caused by ESD are indeed minor. So even though ESD damage is a serious concern in the semiconductor industry, it is important to consider the benefits and place ESD in perspective.

1.2. NATURE OF ESD AND APPROXIMATING MODELS

There are two main types of ESD phenomena in nature: lightning, or ESD over large distances (hundreds of meters) and large voltages (hundreds of kilovolts), and ESD's over small distances (less than a centimeter) and low voltages (approximately a kilovolt). Since in both the air dielectric is broken down, the approximate breakdown fields are the same. In this book we are interested in the methods to protect integrated chips from the ESD threat. The small distance and lower voltage are the relevant phenomena for the semiconductor devices and will be discussed further in this book.

1.2.1. ESD Relevant to Semiconductor Chips

An ESD pulse can be caused by several physical factors, each leading to distinct characteristics depending on the source. These characteristic pulses are broadly grouped into three categories:

- The human body model (HBM) represents a charged human discharging into an integrated circuit (IC). This is defined by the ESD Association Standard 5.1 [HBM1, 1993].
- The charged-device model (CDM) represents die self-charging (field or triboelectric) and then self-discharge. This model is in the process of being defined. Currently there are two methods, one supporting the field-induced method [Renninger, 1991] and the other supporting a socketed method ESD association standard, D5.3 [CDM1, 1997]. The field-induced method is more realistic but less repeatable, whereas the socketed method is less realistic but more repeatable.
- The machine model (MM) represents charging due to machine handling. This is primarily used in Japan and in the automotive industry and is covered under specifications [MM1, 1994].

The three models are illustrated in Figure 1-1.

The CDM stress is the fastest transient and has the maximum peak current, as shown in Table 1-1. The CDM pulse is usually the hardest to protect against. The HBM pulse, which is represented as a double exponential, is a RC-dominated model and has a time constant of ~10–30 ns. The machine model is an LCR model, in which oscillatory currents are observed and which has a time constant of ~15–30 ns. Of these, the CDM zapping will be most destructive to future circuits. For example, Fukuda [1986] observed a 70% decline in CDM robustness for an oxide change from 500 to 300 Å (40% change), whereas the HBM robustness decline was only 30%.

Even so, the best models only approximate reality. In addition to their approximate nature, parasitic inductors and stray capacitors may significantly alter fidelity to the model. The effect of parasitics is most pronounced in CDM

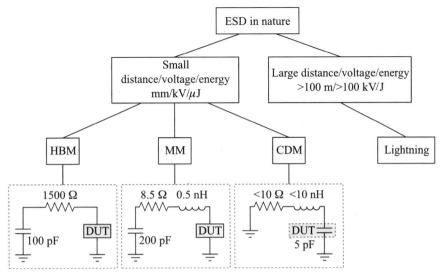

Figure 1-1. Examples of ESD phenomena in nature.

testing, which significantly affects the higher frequency components. To sum up, the test results are only a rough approximation to the ESD event. However, they do allow a consistent basis for comparison of ESD data and provide good repeatability.

1.2.2. ESD-Related Failures

Electrostatic discharge events can lead to failure of poorly protected circuits. These failures are primarily thermally driven. In Figure 1-2, damage to metal caused by CDM stress is shown. The basic phenomenon is for sufficient heat to be generated in a localized volume significantly faster than it can be removed, leading to temperatures in excess of the materials' safe operating limits. The consequences are usually as follows:

TABLE 1-1. Electrical Characteristics of ESD Discharges Showing Peak ESD Current, Rise Time of Leading Edge, and Bandwidth of Energy Content in Transient

Model	I_{peak} (A)	Rise Time (ns)	Bandwidth (MHz)
HBM	1.33	10–30	2.1
MM	3.7–7	15–30	12
CDM	10	1	1100

Source: Amerasekara [1995, p. 18].

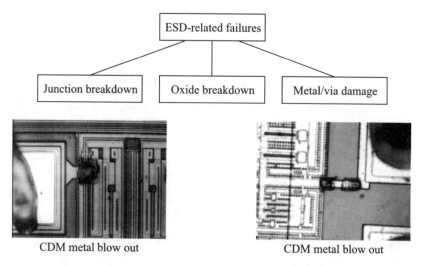

Figure 1-2. The ESD-related failures can be broken down into junction, oxide, or metal/via damage. CDM metal damage is shown.

- Junctions may melt and cross diffuse or lattice damage may occur.
- Oxides may have void formation, vaporization, and filament formation, leading to shorts or opens.
- Metal and contacts may melt and vaporization may occur, leading to shorts and opens.

Future technologies will have shallower junctions, thinner oxides, and smaller line width metals and vias, making them more susceptible to ESD damage. For these reasons a more methodical approach to ESD protection is needed. These issues are discussed in the next section.

1.3. ESD ISSUES IN THE FUTURE

The main ESD issues of the future arise from three reasons, requiring a methodical approach to ESD design:

- increased number of input–output (I/O) and power pins per chip;
- development of increasingly sensitive fabrication processes and an increase in the number of processes to be supported; and
- increased peripheral issues, such as power sequencing, hot insertions, and back powering.

1.3.1. I/O Pin Trends and Sensitivity of Increased Technology

The trend in VLSI today is that the circuit size doubles every 18 months (every year between 1965 and 1970). To make this scaling possible, the lithography and corresponding fabrication processes have to shrink. Figure 1-3 shows that the time dependence of the circuit size increases and lithography features decrease. Slowdowns in the aggressive scaling are occasionally forecast but have not materialized yet, so this trend is very likely to continue for a while.

As the circuit size increases, the number of I/O pins to support these circuits also increases. The required number of I/O pins can be estimated based on the number of gates in the circuit using Rent's rule. This rule is empirically derived and states that

$$N_p = k \times N_{\text{gates}}^{\beta} \tag{1-1}$$

where N_p is the number of I/O pins, k is a constant based on the type of circuit (logic, memory, etc.), N_{gates} is the number of gates, and β is an exponent dependent on the circuit. This function is plotted for various circuit families in Figure 1-4, and the relevant coefficients are tabulated in Table 1-2. It should be noted that logic chips show the fastest growth in I/O pins and the memories show a much smaller rate. Therefore this rule must be modified for parts with significant mix of logic, power, and memory. For example, power devices that traditionally require only a few pins are increasingly becoming "smart" by including more control logic on them. Thus their logic character increases their pin count beyond power devices, but microprocessors that are usually pin intensive are increasingly utilizing on chip cache memory. Thus they

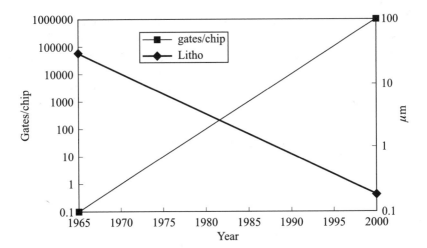

Figure 1-3. Circuit size and lithographic technology trends [Masaki, 1995; Hutcheson, 1997].

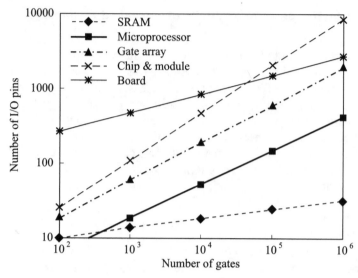

Figure 1-4. Trend in circuit size and I/O pins required for various classes of circuits.

acquire "memory" character, lowering their pin requirements for the corresponding number of circuits they contain.

It is expected that gigascale integration will be achievable by around the year 2000 [Meindl, 1984, 1996]. So it should be expected that the I/O and power pin counts on chips will increase past 1000, in keeping with Rent's rule. Realizing that component reliability is only as good as its weakest pad, the increasing number of pins will certainly create more ESD issues unless a methodical approach is implemented.

To allow this tremendous increase in circuit size, geometries are being aggressively scaled down [Stix, 1997]. These scaledowns also entail decreasing the power supply voltage and gate oxide thickness, increasing doping, and increasing the number of metal layers required [Brinkman 1997; Bakoglu, 1990, p. 26; Yasuda, 1997]. This scaledown has made devices even more sus-

TABLE 1-2 Rent Rule Coefficients for Various Circuit Families

Circuit	β	K_p	Comments
SRAM	0.12	6	Memory
Microprocessor	0.45	0.82	Logic
Gate array	0.5	1.9	Programmable logic
Chip and module	0.63	1.4	Package
Board	0.25	82	System

Source: Bakogolu [1990, p. 416].

ceptible to ESD damage as gate thickness and area and diffused junction depths have decreased [Amerasekara, 1994], requiring less energy and lower voltage to destroy. The energy in the ESD pulse has not decreased. Therefore the ratio of energy required to damage a device to the ESD energy available in a HBM or CDM pulse has been decreasing. One reason is that the volume of the device is decreasing, as shown in Figure 1-5. This ratio does not scale with the current technology and will get even worse with device scaling. This is illustrated in Figure 1-6. Thus, EOS and ESD protection will become increasingly tougher.

Clearly, with increasing I/O counts and stringent protection requirements, the need for an orderly hierarchical design scheme becomes essential.

1.3.2. Increased Number of Technologies

The number of active process generations is increasing. This process of accumulation occurs because process updates are frequently a result of the following:

- improved manufacturing technology,
- optimization of a standard process to support specific designs, and
- retention of "older" technologies.

These can cause significant changes in the operation of breakdown-oriented ESD protection devices, requiring constant monitoring. The increasing number

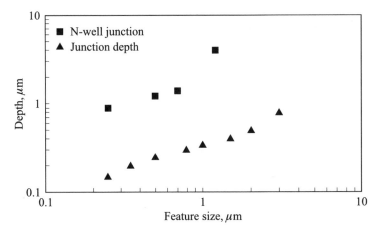

Figure 1-5. Junction depth as a function of feature size. Both decrease simultaneously, thus reducing the device volume in which heat will be generated. In addition, the oxide thickness also scales, lowering the damage voltage.

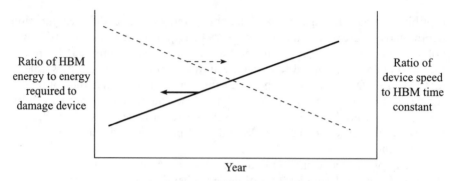

Figure 1-6. Qualitative view of the energy ratio of an HBM pulse to the energy required to damage the device, which is increasing with each process generation. Also shown is the decreasing ratio of the device speed with respect to the HBM time constant.

of processes and their continuous fine adjustments greatly increase both retesting workload and their chances of ESD susceptibility.

Frequently an old process is updated as the learning process allows better manufacturing control. Shown in the Figure 1-7 is an example of gate oxide optimization. Initially, the process may have a large range in thermal oxide thickness. As more experience is gained, this variation is reduced. Once better control is achieved, the target value of the oxide thickness may be reduced such that the new lower limit matches that of the old process. The thinning of the gate oxide allows a higher G_m which improves the process speed. Similar optimizations to other variables occur on a regular basis.

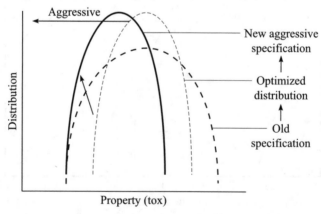

Figure 1-7. Process optimization cycle showing reduction in gate oxide as the process is better understood. In this fashion the old specification is tightened toward a more aggressive and tighter distribution.

Protection devices that are based on breakdown phenomena need to be monitored constantly, as some of these changes may significantly affect their performance. On the other hand, some protection devices have the advantage that when the necessary data or estimates are available, simulations can be used to evaluate device performance and enable any corrections if needed. This avoids costly experiments and saves time. It has also become very popular to support many variants of a base process. These may be targeted specifically at high performance logic or low power logic or different voltages. To complicate matters further, processes have been optically shrunk, and older ones have been hybridized with newer processes to create new optimization variants.

Another subtle influence seen in the IC industry is that old fabrication processes are being retained. This is partly due to voltage scaling. Previously a 5-V process was replaced by another superior 5-V process and the older process was retired. However, now the 5 V has decreased to 3.3 V and this trend will continue. Thus currently each process generation has a unique operating voltage that is different from the other generations. To retain fabrication capability at a nominal voltage, it becomes essential to maintain the older process. This retention phenomenon means that more and more processes need to be supported for ESD, thus straining personnel resources [Johnson, 1993]. A simulation-based ESD methodology can significantly ease the pain of maintaining these processes.

Consider a specific example that combines technology with economics-driven optimization, requiring scaling down of the epi thickness within the same technology. Epi thickness controls the substrate tap resistance (for power delivery) and affects the latchup of the complementary metal–oxide–semiconductor (CMOS) technology. Simultaneously, a thinner epi means less time to grow the epi, resulting in cheaper starting wafers. The thickness of the epi is, however, limited by the out-diffusion from the base P^+ material so it cannot be shrunk indefinitely. At the process development stage, a particular epi thickness was targeted and several silicon-controlled rectifier (SCR) designs were examined and one selected. However, at a later stage, the epi thickness was reduced, and the same SCR failed to function, leading to ESD failures. This led to another cycle of SCR development.

1.3.3. Other Complications

Solving ESD problems can lead to other complications in layout, verification, power-up sequencing, area utilization, and driver/predriver design. With increases in number of processes and volume and types of silicon chips produced, complications can easily multiply and become unmanageable. Therefore a generic and simple ESD solution is highly desirable and is discussed in the next section.

1.4. SOLUTIONS

The authors have found that even with good individual ESD components (such as faster devices), successful full-chip ESD solution is not automatic. It is essential that these components are effectively put together on the chip so that they cover all possible ESD zap conditions. Historically two methods have been implemented:

- *Random Path*: The ESD stress current makes its own path from one pin to another. The weakest link breaks down and conducts heavily. In the process, however, a weak link may suffer permanent damage. These weak spots may change from die to die, and more so from wafer to wafer. It takes several iterations of failure analysis and layout fixes to weed out the weakest links and expose the next weakest link. Finally, one hopes ,the desired ESD performance is reached. This is shown in Figure 1-8.
- *Current Path*: The other method is to provide a designated path by which ESD current can be discharged between any pin combination. The designated path has a very low impedance for an ESD zap, compared to all other parasitic paths. The designated path is determined such that the necessary sizes for the interconnect and protection devices can be calculated. In case of failure, the debug is usually simpler, mainly consisting of identifying the failing device in the path and correcting it. Several such attempts have been described and can be found in the literature [Maene, 1992; Dabral, 1993, 1994; Maloney, 1995, 1996; Merrill, 1993].

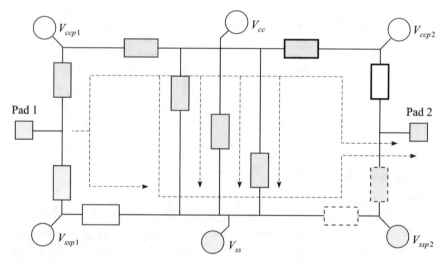

Figure 1-8. Stress current finds its own random path to travel from pad 1 to pad 2 when no designated path is predominant.

The random path is specific not only to the technology but also to the particular chip type (if not the lot type). This defies generalization and each chip type has to be customized.

The current path is more designable and lends itself to more systematization [Merrill 1993, Daral 1993, 1994; Maloney, 1995, 1996]. Figure 1-9 illustrates a generalized path. An ESD zap path between pad 1 and pad 2 is shown. It is essential that the path length (in terms of voltage drop) be sufficiently low such that the oxides and junctions are not damaged. These current paths need component devices to form a working chain. A component device can either be a breakdown-type device or a non-breakdown-type device. These component devices are discussed in the following paragraphs and Chapter 2.

It should be noted that it is difficult to design ESD protection using breakdown devices (BDs) because of their high layout sensitivities and process dependencies. A much easier method to design is using non-breakdown devices (NBDs), which operate near their normal operating regime and are therefore easier to predict and simulate. This book will concentrate on the NBD options.

Two trends that can help NBD protection devices need to be noted:

- An option in the ESD protection of future chips is to use the increasing core capacitance and the higher circuit speeds to participate actively in mitigating ESD stress [Dabral, 1994; Ramaswamy, 1996]. The increasing

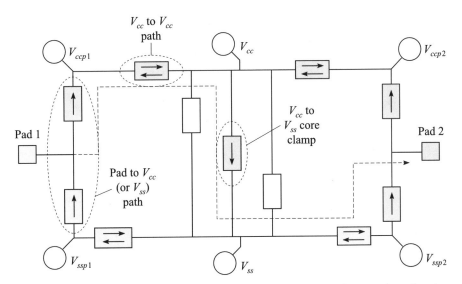

Figure 1-9. Example of preferred ESD current path from one pad to another, showing passage through I/O circuits and power supply clamps. Arrows indicating strongly directional devices are shown. It is also clear how the current path could terminate at any the power supply lines by passing through the elements shown.

circuit size will invariably lead to larger capacitance (due to the devices themselves), and extra on-chip decoupling capacitors will be required to maintain clean power supplies. As the voltage scales down and the current increases, maintaining a low-power-supply ripple will make it essential to have decoupling capacitors in the tens-of-nanofarads range (or larger). These decoupling capacitors help greatly in reduction of ESD voltage stress from building up between V_{cc} and V_{ss}, thus protecting the core circuits.

- With scaling down of feature size, device frequencies increase. Sub-hundred-picosecond gates are obtainable today with speeds rapidly heading towards tens of picoseconds. Compared to these fast gates, the rise times of the fastest ESD pulses (CDM) are becoming comparable (~ 1 ns). In other words, it will become increasingly easier for gates to respond to an ESD pulse.

One must ask: Can we not use these two trends to mitigate some of the enhanced sensitivities that the future ICs will experience?

1.5. OUTLINE OF THE BOOK

Much research has been conducted on ESD over the past 20 years; however, most of it is fragmented, and it has not been compiled into an easy I/O and ESD design reference. Currently books on ESD have concentrated on specifics. For example, data on electrostatics can be found in Greason's [1992] book; SCR design in Diaz's [1994] book; general ESD practices in books by Bhar [1983], Matisoff [1986], and Corp [1990]; and ESD device theory in books by Amerasekara [1995] and Antinone [1986].

Similarly various aspects of VLSI design have been elaborated by several authors [Bakogolu, 1990; Mead, 1980; West, 1985], but integrated information on VLSI, I/O, and ESD is largely absent. Further, some aspects of VLSI design, such as hierarchical and circuit-oriented designs, are very desirable to replicate in the ESD area.

However, there is a large gap in the fragmented research on ESD and design practices in VLSI. This book leverages from past efforts and attempts to bring a simplified design perspective to I/O and ESD design. This will enable a common understanding for both the I/O designer about ESD constraints and the ESD process designer about I/O design requirements. This effort attempts to do the following

- Bridge the gap between I/O, ESD, and VLSI/system design: There is a disconnect between the I/O designer, the process developer, and the ESD circuit development, which this book integrates.

- Examine one ESD methodology that scales with technology: In the future the technology will be increasingly ESD sensitive and the number of I/O pins per chip will increase. This book describes a simple designable methodology that can extend to future technologies without extensive changes or iterations.
- Express device physics in terms of simple electrical circuit models (wherever possible): This allows the practicing designer to converge on an appropriate design and layout for an IC product with ESD protection as an integral consideration.

The ESD methodology will be discussed in Chapter 2. It is based on the earlier discussed "current path" model. Components satisfying these requirements will be identified here. These include coupling diodes and core clamps with the relevant circuit models discussed.

Other key parameters affecting IC protection, such as clamp distribution, package effects, and decoupling capacitance, will be discussed in Chapter 3.

The basics of I/O designs will be examined in Chapter 4. These include simplified buffers, topology, switching noise, and signaling schemes. The timing equations used to determine the speed of the I/O scheme as well as alternatives are also examined. The trend to compensate the I/O buffers, enabling them to be more tolerant to process, temperature, and voltage variations, will be additionally discussed.

The ESD scheme is very sensitive to layout. Layout techniques will be discussed in Chapter 5. Perfectly good schematic designs, or cells, can be useless without attention to the layout of the overall chip ESD protection. The layouts of input, output, and core clamps will be shown, including various options. The metal rules will be derived from the basics and their impact shown.

The ESD components' effect on the I/O performance is discussed in Chapter 6. In addition, the effect of diode coupling noise on the core circuits, potential leakage in I/O buffers due to ESD implementation, and the effect of ESD diodes on I/O transients will also be discussed.

Currently, mixed-voltage designs are becoming more common and have a direct bearing on ESD protection methodology, the I/O output buffer, and predriver designs. This will be discussed in Chapter 7, as well as tolerant buffers and their limitations.

In Chapter 8 methods to generate ESD stress and follow-up with failure analysis will be briefly examined. The methods to construct simple ESD stress tools and ESD characterization methods, both electrical and microscopic, will also be briefly discussed.

Chapter 9 summarizes and concludes the book.

1.6. SUMMARY

Chapter 1 has presented the following:

- A historical perspective was given to help place ESD as a beneficial phenomenon and that the damage to semiconductors is a small price to pay.
- The ESD threat to semiconductors will become severe because of increasing volumes of chips and pins, increasing types of I/O (e.g., mixed-voltage designs), increased number of technologies, and more sensitive devices.
- Three types of ESD model exist: HBM, CDM, and MM. All are approximations to physical reality. The CDM is the severest and usually the toughest to protect against.
- A "random" path model to ESD protection is very hard to design and usually requires several iterations *after* chip fabrication to debug and fix.
- A "current path" methodology is highly designable and has good protection that can be built *before* the chip goes to the fabrication process.
- The scope of this book is to attempt to integrate the best of VLSI practices, ESD, I/O design, and the fabrication process to aspects that lead to a simplified designable and reliable I/O interface.

REFERENCES

[Amerasekara, 1994] A. Amerasekara and C. Duvvury, "The Impact of Technology Scaling on ESD Robustness and Protection Circuit Design," *Proc. EOS/ESD Symp.*, 1994, p. 237.

[Amerasekara, 1995] A. Amerasekara and C. Duvvury, *ESD in Silicon Integrated Circuits*, Wiley, West Sussex, England, 1995, p. 18.

[Antinone, 1986] R. J. Antinone et al., *Electrical Overstress Protection for Electronic Devices*, Noyes, Park Ridge, NJ, 1986.

[Bakoglu, 1990] H. B. Bakoglu, *Circuits, Interconnections and Packaging for VLSI*, Addison-Wesley, Reading, MA, 1990.

[Bhar, 1983] T. N. Bhar and E. J. McMahon, *Electrostatic Discharge Control*, Hayden, NJ, 1983.

[Brinkman, 1997] W. F. Brinkman, "The Transistor: 50 Glorious Years and Where We are Going," *ISSCC*, 1997, p. 22.

[CDM1, 1997] ESD Sensitivity Testing: charged Device Model (CDM)—Component Level, ESD Association Standard, D5.3, 1997.

[Corp, 1990] M. B. Corp, *Zzaap! Taming ESD, RFI, and EMI*, Academic, London, 1990.

[Dabral, 1993] S. Dabral, R. Aslett, and T. Maloney, "Designing On-Chip Power Supply Coupling Diodes for ESD Protection and Noise Immunity," *Proc. EOS/ESD Symp.*, 1993, p. 239.

[Dabral, 1994] S. Dabral, R. Aslett, and T. Maloney, "Core Clamps for Low Voltage Technologies," *Proc. EOS./ESD Symp.*, 1994, p. 141.

[Diaz, 1994] C. H. Diaz, S. M. Kang, and C. Duvvury, *Modeling of Electrical Overstress in Integrated Circuits*, Kluwer Academic, Boston, 1994.

[Fukuda, 1986] Y. Fukuda, S. Ishiguro, and M. Takahara, "ESD Protection Network Evaluation by HBM and CPM (Charged Package Method)," *Proc. EOS/ESD Symp.*, 1986, p. 193.

[Gale, 1994] World of Scientific Discovery, Gale Research Inc., Detroit, 1994, p. 215.

[Greason, 1992] W. D. Greason, *Electrostatic Discharge in Electronics*, Research Studies Press, Taunton, England, 1992.

[HBM1, 1993] ESD Sensitivity Testing: Human Body Model (HBM)—Component Level, ESD Association Standard, S5.1, 1993.

[Hutcheson, 1997] G. D. Hutcheson and J. D. Hutcheson, "Technology and Economics in the Semiconductor Industry," *Sci. Am.*, The Solid State Century, Special Issue, 1997, p. 66.

[Johnson, 1993] C. Johnson, T. J. Maloney, and S. Qawami, "Two Unusual HBM ESD Failure Mechanisms on a Mature CMOS process," *Proc. EOS/ESD Symp.*, 1993, p. 225.

[Lee, 1996] T. H. Lee, "CMOS RF IC Design," Hot Interconnects Tutorial, Aug. 1996, Chapter 1, p. 2.

[Maene, 1992] N. Maene, J. Vandenbroeck, and L. V. D. Bempt, "On Chip Electrostatic Discharge Protections for Inputs, Outputs and Supplies of CMOS Circuits," *Proc. EOS/ESD Symp.*, 1992, p. 228.

[Maloney, 1995] T. Maloney and S. Dabral, "Novel Clamp Circuits for IC Power Supply Protection," *Proc. EOS/ESD Symp.*, 1995, p. 1.

[Maloney, 1996] T. Maloney and S. Dabral, "Novel Clamp Circuits for IC Power Supply Protection," *IEEE Trans. Components, Packaging, Manufacturing Tech.*, Part C, vol. 19, No. 3, July 1996, p. 150.

[Masaki, 1995] A. Masaki, "What Determines the Direction of Packaging Technology?" *Proc. IEEE Multichip Module Conf.* 1995, p. 2.

[Matisoff, 1986] B. S. Matisoff, *Handbook of Electrostatic Discharge Controls*, Van Nostrand Reinhold, New York, 1986.

[McGraw-Hill, 1982] *McGraw-Hill Encyclopedia of Science & Technology*, 5th ed., McGraw-Hill, New York, 1982, p. 167.

[MM1, 1994] ESD Sensitivity Testing: Machine Model (MM)—Component Level, ESD Association Standard, S5.2, 1994.

[Mead, 1980] C. Mead and L. Conway, *Introduction to VLSI Systems*, Addison-Wesley, Reading, MA, 1980.

[Meindl, 1984] J. D. Meindl, "Ultra Large Scale Integration and Beyond," in *Cutting Edge of Technology*, National Academy of Engineering, Washington D.C., 1984, p. 5.

[Meindl, 1996] J. D. Meindl, "Gigascale Integration: Is Sky the Limit?" *IEEE Circuits Devices*, Nov. 1996, p. 19.

[Merrill, 1993] R. Merrill and E. Issaq, "ESD Design Methodology," *Proc. EOS/ESD*, 1993, p. 233.

[Ramaswamy, 1996] S. Ramaswamy, C. Duvvury, A. Amerasekera, V. Reddy, and S. M. Kang, "EOS/ESD Analysis of High-Density Logic Chips," *Proc. EOS/ESD Symp.*, 1996, p. 285.

[Renninger, 1991] R. G. Renninger, "Mechanism of Charged-Device Electrostatic Discharge," *Proc. EOS/ESD Symp.*, 1991, p. 127.

[Stix, 1997] G. Stix, "Towards Point One," *Scientific American*, The Solid State Century, Special Issue, 1997, p. 74.

[van Nostrand, 1995] *Van Nostrand's Scientific Encyclopedia*, 8th ed., Van Nostrand Reinhold, New York, 1995, p. 1134.

[Weste, 1985] N. H. E. Weste and K. Eshraghian, *Principles of CMOS VLSI Design: A Systems Perspective*, Addison-Wesley, Reading MA, 1985.

[Yasuda, 1997] H. Yasuda, "Multimedia Impact on Devices in the 21st Century," *ISSCC*, 1997, p. 28.

CHAPTER 2

ESD PROTECTION METHODOLOGY

2.1. ADDITIONAL ISSUES DUE TO ESD

In the previous chapter, a method based on "current path" to protect against ESD was briefly discussed. In this chapter components needed to create such a path will be identified. The components can be either breakdown (BD) oriented or non-breakdown (NBD) oriented. The breakdown-oriented devices are harder to simulate and characterize compared to the non-breakdown oriented and consequently lead to easier design. Both of these device types will be discussed in this chapter, but design with NBD-oriented devices will be stressed.

The ESD stress not only can destroy the I/O peripheral devices but also can damage weak internal core logic devices. It has been observed that when there is a poorly defined "random" ESD path, multiple failure modes can exist [Duvvury, 1988b]. These failures may also have a tendency to change from die to die, making a consistent solution problematic. Table 2-1 indicates some such failures, mainly the N^+-to-N^+ snapback, the destruction of inverters and clock buffers due to latchup, and the damage to oxides of decoupling capacitors. The main source can be traced to poor core clamping.

Consider a specific example provided by Duvvury [1988b] in the latchup destruction of an inverter. The latchup destruction was identified as occurring between the N^+ guard ring and the N^+ source, as illustrated in Figure 2-1. To prevent destruction of the N^+ regions, the removal of the N^+ guard ring facing the P^+ P-type metal–oxide–semiconductor (PMOS) source and N^+ source was considered. However, this may increase normal latchup sensitivity; thus a partial solution was found by removing only contacts in the N^+ guard ring. This

TABLE 2-1. Selected Issues Identified Due to Poor Core Clamping

Feature/Circuit	Cause/Mechanism	Solution	Reference
N^+-to-N^+ destruction	NPN snapback	Increase space between N^+	Duvvury, 1988b
Inverter latchup leading to destruction	PNPN snapback	Remove N^+ guard ring facing the N device	Duvvury, 1988b
Decoupling capacitance damaged	Oxide breakdown, as core protection thick field oxide (TFO) and did not function	Remove decoupling capacitor	Duvvury, 1988b
Clock buffers damaged	PNPN latchup action	Increase each buffer size to conduct current	Duvvury, 1988b
Clock buffers damaged	Clock buffers faster than the TFO clamp, therefore effectively acting as power (V_{cc} to V_{ss}) clamps but not designed as such	TFO clamp speed improved (reduce channel length), better V_{ss} bussing, and increased drain to gate spacing of clock buffers	Johnson, 1993

effectively lowers the holding voltage of the latched-up SCR, leading to lower heating in the area and resulting in higher robustness. The HBM ESD improved from 600 V to 4 kV in this particular case.

Two points to be noted are that these solutions are very specific to each circuit (not just a technology) and they can only be implemented *after* experimental evidence of damage. The manual effort of failure analysis, design fix, and verification is tedious, complex, and error prone. It would be far simpler to remove the root cause of the issue, that is, to provide a good conduction path between V_{cc} and V_{ss} so parasitic paths are not activated.

Another weak link is the output buffer, in particular the NMOST driver [Maloney, 1985; Polgreen, 1992; Tong, 1996], as verified by the number of issues and correctional steps reported in the literature. This weakness is caused by the low snapback voltage of the parasitic NPN, which leads to second breakdown, which in turn leads to filamentation and subsequently to permanent damage to the transistor. The snapback characteristics of an NPN device are shown in Figure 2-2. During snapback, the current is conducted uniformly by all the fingers of the N-type MOS (NMOS) and the snapback device current scales linearly with the device width. This scaling is possible because during snapback the resistance has a positive temperature coefficient of resistance.

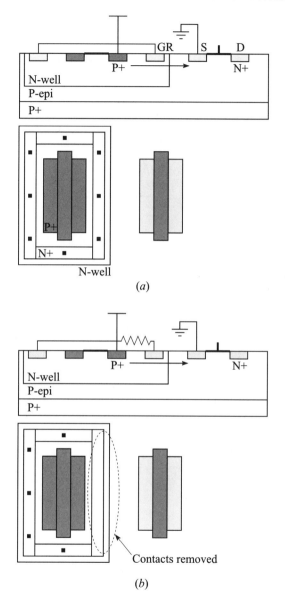

Figure 2-1. Latchup between N and P devices caused by poor clamping of V_{cc} to V_{ss}. (a) One solution is to remove a section of the guard ring contacts between the two devices as shown in (b), leading to complex design rules [Duvvury, 1988].

This implies that if the current in any region increases, that region gets hot, thereby increasing the resistance, which encourages the current to flow elsewhere. However, with the onset of second breakdown, the resistance of the current path has a negative temperature coefficient dependence, encouraging

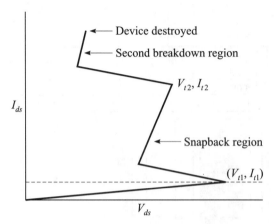

Figure 2-2. A typical NMOS characteristic showing snapback region and second breakdown region.

current "hogging" and filament formation. This nonuniformity in current distribution essentially concentrates the current in certain localized regions (legs) of the NMOS, thus underutilizing the full current potential of the device. The current concentration into a few fingers of a device implies that no matter how many fingers are available only a few will (randomly) turn on and the device ESD current capability will not scale with device size.

To enhance NMOS device self-protection, the current must be made to flow uniformly in all the fingers, which then allows ESD robustness to scale with device size. If the second breakdown voltage is made larger than the snapback voltage, then current will flow in all fingers of the NMOS, which will turn on, and more even current flow will result. Four strategies can be employed to make the snapback voltage comparable to or less than the second breakdown voltage [Polgreen, 1992]:

- Increase the V_{gs} during ESD, which decreases the snapback voltage [Polgreen, 1992]. However, the turn-on must be weak, as a V_{gs} that is too large will degrade the second breakdown current capability.
- Reduce the channel length to decrease the snapback voltage since the NPN β increases. The current capability decreases weakly, but it is compensated for by lowering the snapback voltage. A lower limit is set by the hot-electron degradation of the output driver.
- Forward bias the substrate (slightly), encouraging early snapback. This option is of limited practical use but it is valid if the substrate is free floating.
- Add ballast resistance.

In addition to the uniform current flow option, another alternative is to add a parallel path, which helps to discharge swiftly the major portion of the ESD current. This parallel path drastically reduces the ESD current through the NMOS device, thereby allowing extra robustness. If the parallel path is particularly efficient, sufficient voltage may not build up at the pad, and it may be possible to prevent the NMOS from going into snapback, and then only very weak conduction will result.

The two examples above (latchup of core and output NMOS sensitivity) highlighted the fact that ESD issues are relevant not only to the I/O area but also to any weak spot in the core circuit. In the next section devices that help mitigate ESD stress are discussed.

2.2. DEVICES FOR ESD PROTECTION

The ESD devices that prevent overstress (clamps) are either breakdown devices or non–breakdown devices (see Figure 2-3). Generally, the breakdown-oriented clamps are very area efficient but are considerably harder to design and their operation is harder to predict. This makes them very empirical in design, requiring extensive experimental matrices or iterations. Each of the BDs is briefly discussed here but more attention is given to the NBDs.

2.2.1. Thick-Field-Oxide (TFO) Clamps

A TFO device that can be used as a clamp is shown in Figure 2-4. The device consists of a NMOS transistor with a TFO gate. The gate is tied to the drain, but this has only a weak influence on the drain current. The device primarily

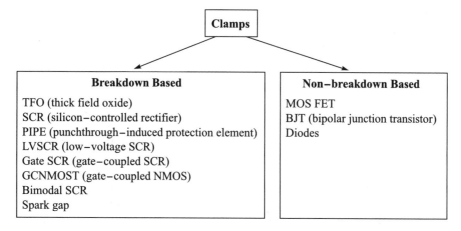

Figure 2-3. ESD protection clamp options by breakdown and non-breakdown conduction devices.

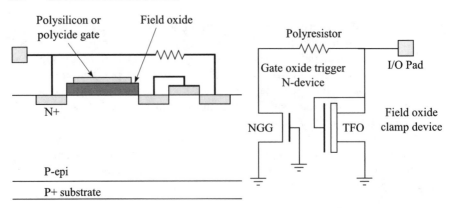

Figure 2-4. Simple TFO clamp. Here the thick-oxide grounded-gate device is the main clamp and an auxiliary thin-gate oxide MOS field-effect transistor in the immediate neighborhood for the TFO device acts as the trigger.

utilizes NPN snapback action to provide clamping. In addition, a thin field oxide device is physically placed close to the TFO device. The drain of the thin oxide device is protected using a resistor [Jaffe, 1990; Johnson, 1993, Amerasekera, 1995, p. 81]. However, the TFO is used to prevent the drain–gate oxide short due to the high ESD voltage at the pad. The trigger NMOS is used in a grounded-gate mode whereas the TFO has its gate tied to the drain.

The action of this clamp is as follows. When high voltage is applied to the pad, the thin-oxide NMOS breaks down and conducts. This breakdown raises the local substrate potential sufficiently to forward bias the source and help initiate the breakdown of the TFO. Once the TFO turns on, it carries the brunt of the ESD current. Since the grounded-gate NMOS is not capable of carrying large currents, the resistor has to be chosen such that the NMOS safe current is never exceeded. This scheme worked well when the diffusions were not silicided. Silicidation has reduced its performance dramatically, primarily due to nonuniform distribution of current in the thick-field device during clamping.

2.2.2. Grounded-Gate NMOS (GGNMOS)

A GGNMOS device is formed by shorting the gate (thin oxide) to the source, as shown in Figure 2-5. The main difference between the TFO device and the GGNMOS is the oxide thickness. The TFO uses a thick field oxide whereas the GGNMOS uses a thin gate oxide. The gate source short ensures that the device is never turned on. During an ESD pulse, the NPN snapback action occurs, which dissipates the energy. Again, the GGNMOS devices worked well when no silicided junctions were used and ballast resistance could be implemented by spacing the drain contacts to a poly edge distance. The ballast action of the N^+ regions provided sufficient protection to the individual fingers and ensured uniform distribution of the current among the fingers. This allowed the device

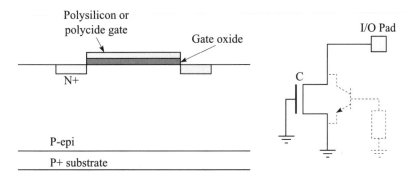

Figure 2-5. A GGNMOS device showing the gate shorted to the source.

ESD clamping capability to scale with the device width. However, with the introduction of the silicided technology, this action has been large eliminated. This has led to nonuniform current distribution to the device. Only a few fingers may take all the ESD current while other fingers do not share the stress, which leads to early failure. The performance of the GGNMOS has correspondingly suffered.

2.2.3. Silicon-Controlled Rectifier (SCR)

The SCR occurs inherently in CMOS technology because of the complementary nature of the devices, which require a well for isolation. However, this term is a misnomer in CMOS technologies because the "control" is very hard to achieve.

A SCR in a CMOS technology is shown in Figure 2-6 [Amerasekera, 1995, p. 75; Duvvury, 1995]. It consists of a PNP transistor formed by the P^+, the N-well of the PMOS, P-epi substrate, and an NPN formed by the N^+ source of the NMOS, the P-epi, and the N-well of the PMOS. In addition, there is the resistance in the N-well and also in the P-epi, both of which are very important factors in the latchup process.

The latchup condition can be modeled by the NPN and NPN pair shown. The total device current I depends on the state of the SCR, whether it has been latched up or is still in the blocking mode. The latchup condition is created where there is sufficient regeneration of current to sustain the flow. This condition can be analyzed with the help of Figure 2-6.

The collector current is a fraction of the emitter current (αI_e) less the current flow due to the reverse-biased junction (I_{co}). When the emitter current is small (due to a blocking junction), I_{co} is a significant fraction of the current and α is small. This can be written as Eq. 2-1 below. This equation applied to the current situation gives the relationship between the collection currents I_{c1} and I_{c2} (Eqs. 2-2 and 2-3). Substituting Eqs. 2-1 and 2-3 into Eq. 2-4, the total device current in the SCR is seen to be dependent on the sum of α_1

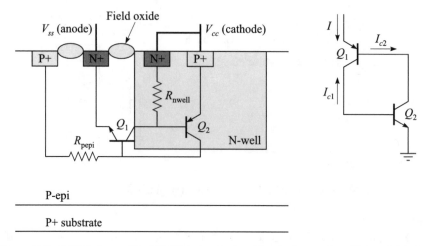

Figure 2-6. A SCR formed in CMOS technologies. The device needs sufficient initiation current to forward bias the bipolar junction transistor before latchup can occur. In this device the N-well to P-epi breakdown provides sufficient current to initiate latchup.

and α_2 [Yang, 1978], as shown in Eq. 2-5. When the sum approaches 1, the current increases dramatically:

$$I_c = \alpha I_e - I_{co} \tag{2-1}$$

$$I_{c1} = -\alpha_1 I_{e1} + I_{co1} \tag{2-2}$$

$$I_{c2} = \alpha_2 I_{e2} - 1_{co2} \tag{2-3}$$

$$I = I_{c1} - I_{c2} = I_{e1} = I_{e2} \tag{2-4}$$

$$I = \frac{I_{co2} - I_{co1}}{1 - \alpha_1 - \alpha_2} = \frac{1_{co}}{1 - (\alpha_1 + \alpha_2)} \tag{2-5}$$

where α_1 and α_2 are the current gain of PNP and NPN transistors, respectively; I_{c1} and I_{c2} are the collector currents and I_{co1} and I_{co2} the reverse-biased collector currents of the NPN and PNP transistors, respectively. It is seen that as the sum of α_1 and α_2 tends to 1, I increases drastically. This is the case when the SCR latches up. This explains the latchup condition, but how does the SCR get initiated into latchup?

When the voltage on the cathode is negative with respect to the anode, the P-epi to the N-well diode is forward biased. Since this P-epi/N-well diode is a wide area that is forward biased during the discharge, no latchup occurs and usually no damage is seen.

However, when the cathode is positive with respect to the anode, the N-well to P-epi junction blocks the voltage. The voltage builds until this junction breaks down. The resulting current flow biases the R_{nwell} and R_{pepi}. With sufficient current flow, the NPN and PNP base and emitter are also forward biased. The result is a positive feedback and the SCR latches up.

A component to this latchup action is due to the capacitive displacement current (between the N-well and P-epi). This displacement current becomes significant when the capacitances are deliberately made large or high-frequency pulses are applied with large dv/dt. The larger current allows higher α, whose sum can add up to 1, in which case latchup occurs. Such a capacitively triggered SCR has also been designed [Ker, 1992].

One major issue with the SCR clamps is the high voltages required to trigger them. These trigger voltages may be in the 30–60-V range. Thus, the circuit is not protected until the trigger voltages are reached, and then the SCR turns on. Also, it takes an SCR finite time to latch up (~ 1 ns) and the SCR can be too slow for some applications [Diaz, 1994; Duvvury, 1995].

2.2.4. Medium-Voltage Triggered SCR (MVTSCR)

The normal SCR had a high trigger voltage, which leads to poor ESD performance. The MVTSCR is a normal SCR that has been adapted by adding a bridging N^+ area between the N-well and P-epi (shown in Figure 2-7) [Duvvury, 1995].

This device uses the fact that N^+/P-epi breakdown is lower than that of the N-well/P-epi. This is a result of the higher electric fields caused by shallower, sharper junctions and heavier doping [Ghandhi 1968, 1977]. Another important factor is the reduced spacing of the N^+ (bridging) and the N^+ anode, which is smaller than the spacing of the N^+ anode and N-well. This leads to a higher α for the NPN transistor. The lower breakdown correspondingly reduces the trigger voltage, which may be ~ 20 V for a 1-μm technology and ~ 10 V for a 0.35 μm technology [Duvvury, 1995].

2.2.5. Low-Voltage Triggered SCR (LVTSCR)

The MVTSCR has a higher trigger voltage compared to the technology requirement. The LVTSCR represents another effort to lower the trigger voltage and speed up the SCR [Duvvury, 1995]. This is shown in Figure 2-8. Starting with an MVTSCR, the field oxide between the bridging N^+ and the anode N^+ is replaced by a gate oxide. The transistor formed is tied up in a grounded-gate fashion.

This configuration reflects the output NMOS transistor, which will break down in the gate-assisted NPN breakdown mode. This transistor reduces the breakdown voltage of the bridging N^+ to anode N^+. This current raises the P-substrate potential, thereby allowing the NPN transistor to fire. The resulting device is faster and has a lower trigger voltage, which may be in the 8-V range for a 0.35-μm technology. The LVTSCR behavior is similar to the output NMOS snapback. It takes little time for the LVTSCR to latch up after the

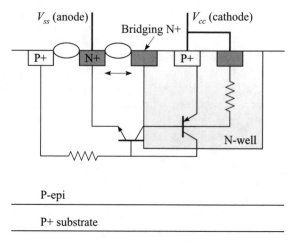

Figure 2-7. A MVTSCR is made by adding an N$^+$ diffusion, bridging the N-well to the P-epi substrate. The N$^+$ bridging causes earlier breakdown, resulting in lower trigger voltages.

snapback event. Therefore, if both the SCR and the output transistor are the same lengths, then the SCR should be expected to latch up later than the output NMOS device. One method would be to reduce the device length of the SCR's trigger NMOS such that it triggers earlier than the output NMOS, thus providing a better protection for the output device. However, if placement of the thin gate oxide is not carefully designed, failures can occur. This has indeed been observed [Duvvury, 1995].

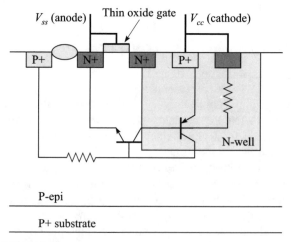

Figure 2-8. A LVTSCR showing the thin-gate-oxide NMOS added to create faster and lower turn-on voltage device.

The previous sections have described a fixed voltage triggering "dumb" SCR designs, but a new breed of programmable "smart" SCRs have been designed. The smart SCRs are able to distinguish their environment and behave appropriately. These are described in the next section.

2.2.6. Bimodal SCR

The "bimodal SCR" is a general term used here to describe the SCR design [Diaz, 1994; Croft, 1994] that functions at two triggering levels:

- high trigger voltage when the circuit is normally "on," or plugged into a printed circuit board, and
- low-voltage operation when the IC is "off," or not plugged into the printed circuit board.

In the first case, shorts are provided externally on the board, as shown in Figure 2-9. This allows the SCR to determine whether it is in a board or not. The circuit operation is as follows. When the pins are not shorted and there is an ESD pulse at the pad, the dv/dt displacement current and the leakage current are forced to flow through the bipolar junction transistors (BJTs) and the substrate and N-well. The only paths are through the NPN and PNP transistors. With sufficient current flow through their emitter–base regions, SCR regeneration occurs. This action is the same as discussed earlier in Section 2.2.3. However, when external shorts are provided, displacement and leakage currents have a low-impedance alternative path, and they do not flow through the emitter–base regions of the NPN and PNP transistors. The circuit appears as a single reverse-biased junction and no regeneration is possible to cause latchup.

In this clamp circuit, two additional pins are needed to program the trigger voltage mode. In several chips that use separate peripheral and core supplies, these two pins can be the peripheral supply pins. However, these additional pins need to be ESD protected. To protect each pin with respect to the other three, three additional SCRs are needed (shown in Figure 2-10). These SCRs provide paths from each pin to another, either directly or in combination with parasitic diodes. Using such a scheme, Croft [1994] has improved an opamp circuit protection from 500 to 4000 V. The trigger setting for this example is 2 V for low and 145 V for high trigger levels. Both are selectable using the board short method.

In another variant of this programmable triggering, the triggering time constant of the clamps was altered [Croft, 1996] using an external board-mounted resistor. In this scheme the RC time constant for an unplugged component was set high so the clamp (Darlington NPN between V_{cc} and V_{ss}) is turned on at a low voltage and protects the internal circuits. When the component is plugged into the board, the small resistance on the board decreased

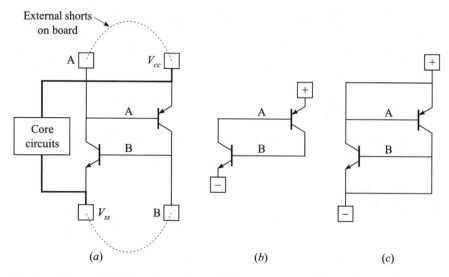

Figure 2-9. Bimodal SCR whose trigger voltage is selected by (*a*) providing external shorts on board. An SCR circuit (*b*) when not shorted and (*c*) when shorted showing the disabled emitter base of the NPN and PNP transistors.

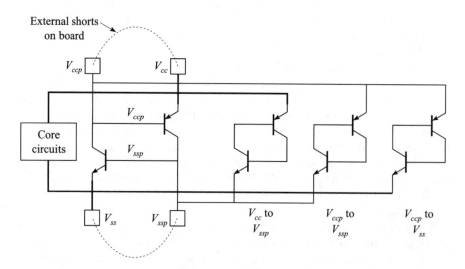

Figure 2-10. Typical arrangement of SCR to protect all the pins. Three additional SCRs have been added to the V_{cc} to V_{ss} protection SCR. Here the peripheral supply has been used to program the trigger voltage [Croft, 1994].

the *RC* time constant to a degree where the protection did not turn on, thereby preventing leakage or latchup.

As a word of caution, the external short is not ideal [Lin, 1994]; the parasitics associated with it are shown in Figure 2-11. These parasitics are due to the capacitance of the bond pads (~3 pF), the inductances of wire bonds (~3 nH), and the short transmission line on the printed circuit board (PCB). Consider a *di/dt* of 160 mA in 1 ns passing through a 6-nH inductor, which will generate a 1-V drop across the transistor base–emitter Q_2 effectively forward biasing it. Thus for such sharp transients (as could occur during a CDM pulse), the external short may not function as a short and could potentially latch up an active circuit. Therefore, care must be taken to use adjacent wire bonds and a very small (and wide) trace on the PCB for shorting purposes. The adjacent wire bond strategy decreases the wire bond inductance since the outgoing and return current paths will enclose a small area.

The last approach needed external pins to program the SCR trigger points. Circuit techniques can also be used to provide the similar bimodal functionality, as shown in Figure 2-12. Consider the circuit developed by Diaz and Motley [Diaz, 1994]. This circuit is similar to the LVTSCR, only instead of being grounded the gate is tied to a control. The gate is controlled using a

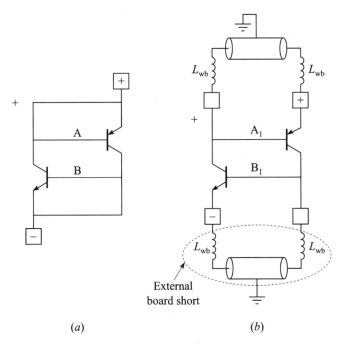

(a)   (b)

Figure 2-11. An electrical model for an external board short utilized for the SCR programming.

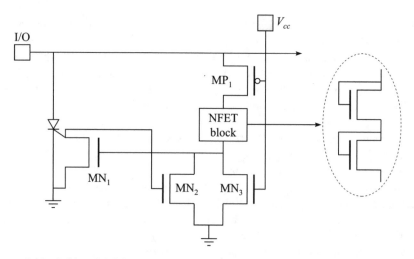

Figure 2-12. A bimodal SCR that can be programmed using the power supply condition of the IC [Diaz, 1994].

simple control circuit that senses the ESD pulse and then triggers using part of the ESD energy.

The LVTSCR can trigger only when there is no V_{dd} applied, as would be the case when the chip is not on a powered-up board. If a V_{dd} supply is present, the PMOS MP_1 is turned off and the NMOS MN_3 solidly grounds the trigger node, blocking any triggering. To tailor the trigger point, the turn-on of the NMOS MN_1 must be controlled. This is done using the diode-connected N-type field effect transistor (NFET) block. By varying the number of cascaded NMOS transistors, variable-trigger-voltage capability is obtained. However, the cascading does not lead to integral multiples of V_{tn} blocking but rather a fraction of that [Diaz, 1994].

This SCR may not function well for the CDM. In the example provided by Diaz [1994], the SCR fails when small rise time pulses are applied $(dv/dt > 34\,\mathrm{V/ns})$ but they are well protected when the larger rise time pulses are applied $(dv/dt \sim 12\,\mathrm{V/ns})$. From this it can be estimated that the LVTSCR may be able to respond only in several nanosecond time frames whereas the technology $(0.6\,\mu\mathrm{m},\ \mathrm{CMOS})$ is capable of $\sim 200\,\mathrm{ps}$ gates. This implies that when protecting an NMOS, this device may be too slow. The PMOS is the most sensitive element in this circuit. In terms of layout, the spacing sensitivity of N^+ and P^+ is important, as in any SCR. These circuits also show a spread of trigger voltages (as high as 20% around a mean value).

2.2.7. Gate-Coupled NMOS (GCNMOS)

It has been observed that the NMOS transistors conduct current nonuniformly when silicided technology is used. This nonuniformity is explained with the help of Figure 2-13. Consider an NMOS finger that has snapped back at V_{t1} and has a current of I_{t1} due to external applied voltage. Now the voltage in the drain region tries to decrease to a minimum point of V_{sb} at snapback. However, the external voltage (due to ESD) does not allow this to happen, as it is directly connected to the drain node, so to meet the external voltage, the current in this snapped-back finger has to increase. If the increased current cannot cause the drain voltage to decrease below V_{t2} (perhaps by discharging the external charged source), then the device passes into the second breakdown region. Here a thermal runaway process starts, and in this region the device experiences a negative temperature coefficient of resistance, which implies that the current will keep increasing for the same applied voltage. This current increase will ultimately cause sufficient heating such that the local area melts

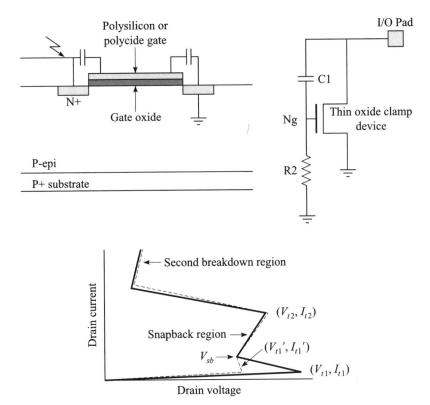

Figure 2-13. Simple GCNMOS structure. The capacitor and resistor combination couples enough charge to help turn on the NMOS very weakly, ensuring lower snapback voltage (V_{t1}') and allowing all the fingers to turn on.

and is destroyed. If the second breakdown voltage V_{t2} is smaller than the snapback voltage V_{t1}, then the conducting finger will be destroyed even before other fingers snap back and share some ESD stress. This mechanism shows that if the drain is directly tied to the I/O pad, a finger that broke down earlier will continue to carry increasing amounts of current leading to its destruction. Increasing the number of fingers will have no impact on the ESD robustness in such devices. A ballast resistance allows separation of the pad and the NMOS transistor drain such that the drain voltage is significantly modulated by the current flowing into the drain. This allows other fingers to snap back and more uniform ESD current distribution among the fingers.

Another method to rectify this nonuniform breakdown is to use the GCNMOS design. A GCNMOS is formed when a capacitor is used to couple a fraction of the ESD charge to the gate of the NMOS, as shown in Figure 2-13. This is done using a high-pass filter circuit consisting of a capacitor (C_2) and a resistor (R_2) [Amerasekera 1995, p. 69; Duvvury 1995; Ker, 1997]. These are deliberately drawn capacitors in addition to the parasitic capacitance already existing. At the initial stage of an ESD pulse, enough charge is coupled into the resistor such that it weakly turns on the NMOS device.

By allowing the gate to turn on weakly, the snapback voltage is lowered, shown in Figure 2-13 as (V_{t1}', I_{t1}'). If the snapback voltage is lowered sufficiently ($V_{t1} < V_{t2}$), then other fingers in an NMOS device will also snap back before any one finger goes into a second breakdown. This allows uniform current distribution, and ESD performance then scales with device size.

For a GCNMOS design, a target in terms of ESD voltage and the desired gate voltage can be chosen. Further, by knowing the voltage appropriate charge coupling capacitor, resistance can be calculated. The second breakdown current at this gate voltage must also be known (i.e., experimental or simulation). The device width can then be calculated based on the required tolerance of the ESD zap current and the second breakdown current I_{t2}.

There are a few issues with this circuit. One needs to experiment, simulate, or estimate I_{t2} before the device is sized and drawn. Changing parasitics will affect coupling, so it needs to be redesigned for each new technology. This scheme certainly enhances the NMOS performance by ensuring all devices conduct current, but it is still based on the breakdown phenomenon. Further, this action works well for HBM-type pulses. However, for CDM pulses its effectiveness is reduced [Duvvury, 1995].

In the previous sections the SCR and GCNMOS were discussed. Table 2-2 summarizes the primary characteristics of these clamps.

2.2.8. Punchthrough-Induced Protection Element (PIPE)

The PIPE [Kwon, 1995] is a clamp that works similar to a diode in one direction and as a punchthrough device in the other. The construction of the device is shown in Figure 2-14. It consists of an N^+/P-well diode between V_{ss} and the I/O pad and an NPN transistor between the I/O pad and V_{cc}.

TABLE 2-2. Comparison of ESD Performance, Damage Mechanism, Solutions for MOSFETs, and SCR Devices

	GGNMOS	GCNMOS	MVTSCR	LVTSCR
Trigger voltage	—	—	20 V at 1 μm, 10 V at 0.35 μm	15 V at 1 μm, 8 V at 0.35 μm
Nonsilicided HBM current capability	10 V/μm	10 V/μm	40 V/μm	40 V/μm
CDM	$+V_e$ better, with pad to V_{ccq} diode more robust	$+V_e$ better, with pad to V_{ccq} diode more robust; +1000 V, −600 V	+1500 V, <−500 V	+500 V, > −1000 V
Silicided CDM without diode	—	−1 V/μm, +2 V/μm	—	—
Silicided CDM with diode	—	−2 V/μm	—	—
Damage mechanism	Melting of source contact; melt filament between source and drain	Melting of source contact; melt filament between source and drain	+CDM more robust than LVTSCR (3×); − CDM less robust than LVTSCR because of slow response of SCR (3×)	+CDM shows lateral NPN action, resulting in damaged thin oxide; −CDM shows anode-to-cathode damage
Fix	Increased metal widths, area, and number of contacts and lower inductance	Increased metal widths, area, and number of contacts and lower inductance	Increased device width and improved diode action	Increased device width and improved diode action
Preferred path	—	—	Vertical NPN	Lateral NPN

Source: Duvvury [1995].

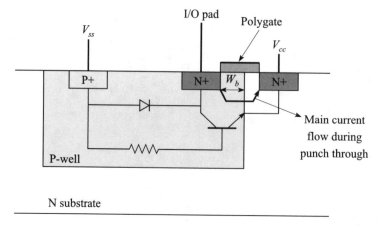

Figure 2-14. Cross section of PIPE device showing placement of critical poly gate for setting up the right punchthrough voltage.

When the I/O pad is positive with respect to V_{cc}, the P-well is reverse biased and the depletion layer extends toward the P-well/N-substrate boundary. Once the depletion layer touches the P-well/N-substrate depletion layer, a low-impedance path is created between the I/O pad and the V_{cc}, resulting in current flow that then biases the NPN. A sharp snapback-like lowering in voltage is observed. The poly gate length is used to control the punchthrough voltage, and with larger gate length, the punchthrough voltage increases. However, this gate length also affects the ESD performance as the punchthrough voltage is modulated.

The gate length is not a completely free variable, as it is needed to design a punchthrough voltage. There are additional sensitivities to the P-well doping and channel implant that control the punchthrough voltage. With increasing P-well doping and channel implant the punchthrough voltage increases. However, with increasing P-well doping ESD performance was found to improve, whereas with increasing channel implant ESD performance was degraded.

It can be speculated that P-well doping reduces the P-well/N-substrate boundary to the N^+ (tied to V_{cc}) distance. This reduction in distance reduces the resistance in the ESD current path, which is not at the surface but slightly under the surface (as shown in Figure 2-14). This reduction in resistance offsets the increase in punchthrough voltage, and an overall improvement in ESD performance is seen. On the other hand, the channel implant only increases the punchthrough voltage, leaving the DC resistance in the current path unchanged. In addition, this current spreading distributes the heat generation, thereby providing the PIPE with ESD robustness.

The main advantages of a PIPE are small area and reasonable ESD performance. However, breakdown characteristics, P-well sensitivities, gate length effects on ESD performance, and punchthrough voltage need to be traded off.

2.2.9. Spark Gap

Spark gaps have been used in IC protections [Bahr, 1983] but are becoming increasingly unviable in IC technologies. These devices rely on electron emission under a strong electric field [Wallash, 1994]. This electron emission occurs in two modes: the Paschen mode, when the gap is larger than about 25 µm (low electric fields), and the field emission mode, when the gap is smaller than 10 µm (high electric fields). In the Paschen mode, the breakdown voltage is high (> 650 V), which makes such devices unsuitable as IC clamps. However, in the field emission mode much lower voltages can create a breakdown.

Their impracticality is primarily due to their larger trigger voltage, their slow response, and the reliability of the gap both before and after zaps. As an example, when spark gaps are used to protect magnetic heads, it has been reported that a gap less than 3 µm results in shorts after zapping (Figure 2-15). Thus gaps greater than 3 µm are required, which makes the trigger voltage greater than 500 V. In addition, these devices will show wear-out after multiple zapping. One can argue that in current silicon technology with finer lithography, gaps can be made in micrometer size, thereby reducing the trigger voltage and reducing shorts. However, the wear-out mechanism will still exist and needs careful design.

Another issue reported is that with charge sharing between the gap and the discharging body a voltage sufficient to trigger this device may not be reached, in which case the gap will fail to trigger. The added device capacitance also slows the ignition time at a rate of 135 ns/pF. Thus for ICs having large capacitance, the gap will be too slow and may not even ignite, leaving a

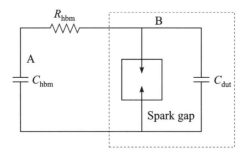

Figure 2-15. Schematic of a spark gap together with the device capacitance. The device capacitance shares charge from the HBM capacitor and may lower the voltage sufficiently that the gap does not ignite [Wallash, 1994].

possible protection hole. Clearly this device is not suitable for IC clamps unless additional enhancements can be made.

2.2.10. Zener Clamps

Zener clamps are simply diodes that avalanche at relatively low voltage so that they are able to protect other circuitry. Zener diodes are created by placement of heavily doped (even degenerately doped) junctions adjacent to each other. In a regular CMOS or bipolar MOS (BiCMOS) process such junctions are hard to create in a standard form.

Corsi [1993] provides an example of zener implementation using the most heavily doped junctions available, the emitter and the base. The circuit is shown in Figure 2-16. Here, one NPN transistor (Q_2) is broken down deliberately during an ESD event. The breakdown voltage is BV_{ebo} and current flows through resistor R_1 to ground. The BV_{ebo} has to be preferably lower or about the same as the snapback voltage for the NMOS transistors. Usually the emitter–base breakdown of Q_2 has a small volume of silicon to dissipate the ESD energy, creating large power density. So dissipating all energy in the transistor Q_2 is not practical. A vertical NPN transistor with a vertical current flow and large dissipation volume is preferred. Resistor R_1 can be used to bias the NPN Q_1, which provides the main ESD discharge path. For the case where the pad is negative with respect to the pad, diode D_2 provides the main conduction path.

In BiCMOS technology the bipolar collector current flows vertically and NPN has significant gains. Therefore this scheme is more applicable to BiCMOS technologies. In a regular CMOS technology the NPN is usually absent.

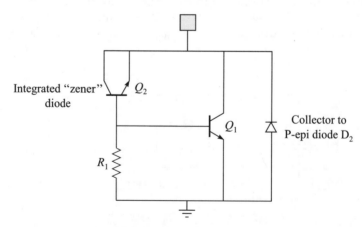

Figure 2-16. Zener action-based clamp. Here the zener diode and the resistor are used to sense an ESD transient and the NPN transistor is used to shunt the majority of the current.

2.2.11. Double-Implant Field Inversion Device in Well (DIFIDW)

This device is constructed using a TFO as the gate dielectric and a deep N-well as the source and drain [Lin, 1984], illustrated in Figure 2-17. It has a high threshold voltage, primarily because of the thick oxide, and in this example there is 23 V for a 10,000-Å oxide and 14 V for a 6000-Å oxide. The deep N-well source and drain were used as protection against metal spiking, which can occur in poorly protected shallow N^+ junctions [Maloney, 1990]. This gave an additional level of safety for direct aluminum metallization onto silicon, which did not have the benefit of contact barriers.

This device may be diode connected to form a clamp that triggers on only at high voltage and it is off at normal operating voltage. Even though this device operates in a conduction mode, it may suffer snapback. Here the thick gate oxide helps prevent oxide damage. The DIFIDW is different from a TFO primarily in its use of an N-well as the source and drain. This option was valid when little aluminum-to-silicon contact barriers were implemented. With barriers placed in the contacts, the necessity to have a deep N-well (to prevent aluminum spiking) has diminished and TFO devices are sufficient.

In the last section, breakdown-oriented clamps were discussed. These devices need considerable amount of design usually followed by several trials. This may include defining the device in three dimensions and then simulating their performance on a computer. Thus considerable information on the device can be obtained even before devices are laid out in silicon. This technique is discussed in the next section.

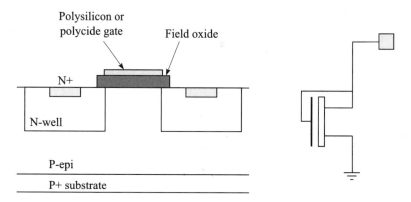

Figure 2-17. Simplified cross-sectional view and circuit schematic of a DIFIDW device used as an ESD clamp.

2.3. ELECTROTHERMAL SIMULATION

Electrothermal simulation tools can be utilized to optimize and characterize the device performance before mapping them to silicon. These tools use finite-element methods and solve the basic semiconductor equations and include the effect of heat generation and flow [Krabbenborg, 1991; Mayaram, 1991]. The models are extremely complex and based on such assumptions as no filamentation, uniformity, and no hot spots. They are extremely useful to analyze the device physics. However, they are very cumbersome for engineering ESD protection circuits.

For example, an electrothermal study can help validate an assumption for second breakdown, which has been that it occurs when the intrinsic silicon electron density (n_i) equals the doping density of the silicon. However, it is very difficult to verify this experimentally as junction temperatures cannot be easily measured in a real device. It is further complicated by nonuniform doping. For second breakdown the heat generation sites have to be identified, and one needs to evaluate where the voltage is being dropped. For example, in a resistive material, the peak temperatures may be away from the junction (depletion layer) depending on the J·E product. However, in low-resistivity materials, the major voltage drop is across the depletion layer, which is the location of peak temperature. Interactions such as nonuniform doping and heat generation defy analytical solutions, and simulation is necessary. These simulations can help predict where the second breakdown will be initiated and help optimize a device.

It should be pointed out that if one were to base all protection on forward-biased junctions, which dissipate less power/unit volume, then the second breakdown phenomena is largely avoided and these isothermal tools are of limited significance.

2.4. NON–BREAKDOWN DEVICES

Non–breakdown devices are primarily diodes, MOSFET, and bipolar junction transistors (BJTs). The standard models and physics of these devices are well understood and several standard texts [Yang, 1978; Tsividis, 1987; Pierret, 1990; Sze, 1981] discuss the small current and voltage regime characteristics thoroughly. The large voltage and current regime characteristics are different [Ghandhi, 1977]. Both of these regimes are of significance, as in the off state the ESD clamp follows the low current–low voltage behavior and in the on state it follows the high voltage–high current behavior.

2.4.1. Diodes

Diodes are the simplest and earliest solid-state semiconductor devices. For ESD purposes they are very efficient, forgiving, and robust. It is therefore critical to understand their behavior from an ESD perspective in an IC environment.

Diode characteristics are fairly simple to approximate, model, and simulate. The diode has four primary regions of operation: reverse bias, weak forward bias, forward bias, and strong forward bias (see Figure 2-18). An ideal diode will have current in the forward bias as described by the equation

$$I = I_0 \left(e^{V/V_t} - 1 \right) \qquad (2\text{-}6)$$

where I_0 is the reverse saturation current, V is the forward bias, and $V_t = kT/q$ (26 mV at room temperature). So the voltage supported by an ideal diode is given by

$$V_f = V_t \times \ln\left(\frac{I}{I_0} + 1 \right) + IR_{\text{diode}} \qquad (2\text{-}7)$$

The +1 term can be neglected even for relatively weak forward bias, while for small current densities, the IR drop is small and can be neglected. In the ESD context, the low IR drop is in the leakage regime. During an ESD even several amperes of current will flow and the IR component is a significant factor and needs to be considered.

For an ideal diode, the forward voltage decreases as the temperature is increased:

$$V_f(T_1) = nE_{g0} + \left(\frac{T_1}{T_0} \right) [V_f(T_0) - nE_{g0}] \qquad (2\text{-}8)$$

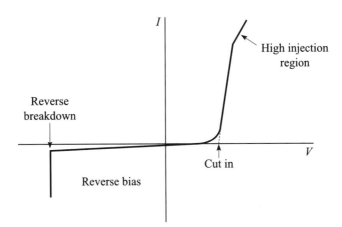

Figure 2-18. Operating regions of a diode showing the reverse-biased, weak forward-biased, forward-biased, and strongly forward-biased regions.

where $E_{g0} = 1.206$ V and n is the diode ideality factor; T_1 and T_0 are the new and original temperatures.

These properties hold for an ideal diode. The diodes are affected significantly by the type of process used. The CMOS process utilizes a well to provide the bulk for the complementary transistor. This well formulation may also have an effect on the diodes, which is discussed next.

2.4.1.1. Cascaded Diodes
In N-well CMOS technology a diode is actually a PNP transistor. This is illustrated in Figure 2-19, which shows a chain of four serially connected diodes. This parasitic nature provides the diode some interesting properties that can be beneficial as well as detrimental.

The single-transistor element is shown in Figure 2-20. Instead of a two-terminal device, the CMOS diodes formed in a well are really three terminal devices. In some technologies, the transistor action can be very pronounced. The consequences of chaining these devices then becomes significant.

If ideal diodes are cascaded, the cut-in voltage increases linearly with the number of diodes in series. However, for a diode chain constructed using PNP transistors, the linear increase in cut-in voltage with the number of diodes in a chain is lost. The PNP action allows some fraction of the emitter current to sink into the substrate; thus there is less current in the next stage. This lowering of current reduces the cut-in voltage of the next stage. This action is repeated at each stage. Thus the sum of voltages in a cascaded PNP is smaller than would be the case for cascaded diodes. The cut-in voltage of identically cascaded diodes is given by

$$\ln\frac{I_1}{I_S} = \frac{qV_1}{nkT}$$

$$\ln\frac{I_2}{I_S} = \frac{qV_2}{nkT} = \ln\frac{I_1}{(\beta+1)I_S} = \ln\frac{I_1}{I_S} - \ln(\beta+1)$$

(2-9a)

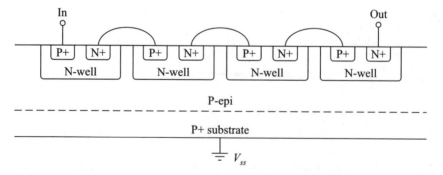

Figure 2-19. Diode chain cross section showing the parasitic PNP devices.

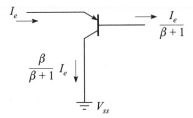

Figure 2-20. Current gain effects in the PNP bipolar transistor formed by a diode stage.

so that

$$V_2 = V_1 - \frac{nkT}{q}\ln(\beta + 1) \quad \text{or} \quad V_2 = V_1 - \ln(10)\frac{nkT}{q}\log(\beta + 1) \qquad (2\text{-}9b)$$

Now let $V_0 = \ln(10)(nkT/q)$, which is 60 mV for an ideal diode at room temperature. The analysis of Eq. 2-9 is applied to multiple stages to give a loss of an additional $V_0 \log(\beta + 1)$ at each stage, resulting in a total voltage V_t of a string of m identical diodes at current I_1:

$$V_t = mV_1 - V_0 \log(\beta + 1)(\tfrac{1}{2}m(m-1)) \qquad (2\text{-}10)$$

where m is the number of cascaded diodes. If the diode chain does not consist of equally sized diodes, then the voltage can be calculated using Eq. 2-11, which accounts for the changes in the area:

$$V_t = mV_1 - V_0 \left(\sum_{i=2}^{m} \log[A_i(\beta_{i-1} + 1)^{m-i+1}] \right) \qquad (2\text{-}11)$$

The effect of the parasitic action on a string of identical diodes is shown in Figure 2-21. The transistor action prevents linear buildup of forward voltage. Increased temperature reduces the forward voltage even further. Again this property can have desirable and detrimental effects (see Section 2.6).

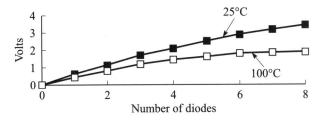

Figure 2-21. Diode string turn-on voltage for two temperatures ($\beta = 6$).

It is seen that, due to the multiple cascading action, at each step a fraction of the current is passed on to the next. So after a few stages only a small fraction of the current remains. This can also be observed from a control point of view where a small current at the end of the chain can be amplified into a larger current at the first stage of the chain. This cascade acts as one large Darlington transistor and its property will be utilized (discussed in later sections.)

The quantity β is current dependent. A good transistor should have a very high minority carrier injection into the base from the emitter (I_{BE}), a very small recombination current (I_R), and a very small minority carrier injection from the base into the emitter (I_{EB}). Roughly, the ratio of I_{BE} to the sum of I_{EB} and I_R gives β. For very light forward biases β is small because the diffusion current I_{BE} is small and the recombination current I_R is comparable. This lowers β. However, at medium biases, the recombination current I_R is overwhelmed by the diffusion current I_{BE} from the base, and β peaks. As the forward bias increases, the charge modulation of the base increases the charge injection from the emitter into the base; therefore I_{EB} increases. This again results in a decrease of β [Ghandhi, 1968, p. 324]. This variation of β as a function of collector current is shown in Figure 2-22. The attenuation of β with collector current is termed the Webster effect [Webster, 1954; Maloney, 1995]. A similar effect is also seen as a function of the emitter current. The emitter current and β can be related as shown in the equation

$$\beta \propto \frac{1}{J_e^n} \tag{2-12}$$

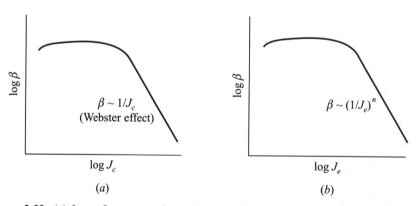

(a)   (b)

Figure 2-22. (a) Log of current gain vs. log of collector current density for a typical diode string PNP transistor. A similar curve results for emitter current density. (b) Log of current gain vs. log of emitter current density for a typical diode string PNP transistor allows simplified modeling of the diode string performance. The coefficient n is measured for each process.

TABLE 2-3: Measured Model Parameters for Various Processes

Process	β (Low J)	β (High J)
P1	56	$(3.71/J_e)^{0.74}$
P2	7	$(1.15/J_e)^{0.509}$

Note: Generally, the high current density (mA/μm) of β can be extrapolated to low currents until it is limited by the low current density (maximum).

where J_e is the emitter current density (in milliamperes per micrometer) and n is a fitting exponent. Some measured data shown in Table 2-3 indicate that J_e is in milliamperes per micrometer and n is in the 0.5–0.75 range.

Using the current-dependent β, a simple Ebers–Moll BJT model can be constructed and is shown in Figure 2-23. Here, G_c and G_b are the conductance/length of the emitter and base regions and L is the length of the diode fingers used. This model is simple enough that it can be entered into a spreadsheet to perform current–voltage calculations for small and large currents with reasonable accuracy. This model has been used and will be referenced in several places.

2.4.2. MOSFETs

Conventionally MOS devices have been the weakest link under ESD stress [Bahr, 1983]. So their direct use to mitigate ESD stress is still counterintuitive. However, several researchers have demonstrated that, when biased correctly, MOSFETs can indeed self-protect or protect other components in a circuit. An early MOS input protection circuit is shown in Figure 2-24.

The MOS device can also be operated as a snapback device, as has been done in the TFO and in grounded-gate devices. These are hard to simulate and predict. For the MOS to operate in a non-breakdown mode, the drain voltage has to be kept under the snapback voltage. For this to happen, the MOS device

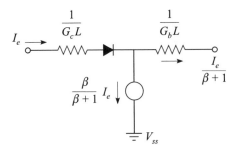

Figure 2-23. High-current regime model for PNP transistor stage.

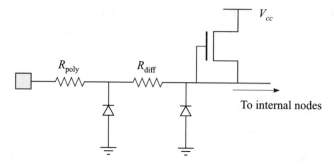

Figure 2-24. An early NMOS clamp used at an input node. The NMOS turns on whenever the node exceeds $(V_{cc} + V_{tn})$ voltage [Hulett, 1981].

has to be large with low on resistance and the gate drive should be quicker $(V_g > V_t$ before breakdown) than the voltage buildup.

If the MOS is kept out of the snapback region, then normal MOSFET design equations and theory can be used. To recapitulate, the I–V curves are shown in Figure 2-25. The current in the saturation region $(V_{ds} > |V_{gs} - V_{tn}|)$ is given by the equation

$$I_{ds} = \frac{Z\mu C_0}{2L}(V_{gs} - V_t)^2 \tag{2-13}$$

and in the linear region $(V_{ds} < |V_{gs} - V_{tn}|)$ by

$$I_{ds} = \frac{Z\mu C_0}{L}\left[(V_{gs} - V_t)V_{ds} - \frac{V_{ds}^2}{2}\right] \tag{2-14}$$

where Z/L is the device width-to-length ratio, C_0 is the oxide capacitance per unit area, V_{gs} and V_{ds} are the gate-to-source and drain-to-source voltages, and

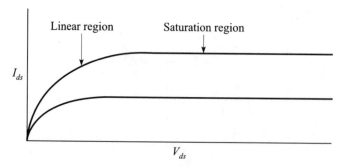

Figure 2-25. The I–V characteristic of a NMOSFET device.

V_t is the threshold voltage of the MOSFET. These are simplistic design equations that ignore the effect of short channel effects, the drain modulation of current, and the velocity saturation effects. So for greater MOSFET details other device physics books should be consulted [Yang, 1978; Sze, 1981; Tsividis, 1987; Pierret, 1990].

2.5. CURRENT PATH CONSTRUCTION

In Sections 2.2 and 2.3 the basic devices required to create a current path for ESD protection were discussed. These include breakdown devices such as SCRs and non-breakdown devices such as diodes and MOSFETs. These can now be applied to the creation of a current path. For this purpose, consider the path shown in Figure 2-26. This path guarantees that any ESD zap will have a designed low-impedance path to discharge and no major portion of the current will flow through the core circuits.

The current path can be broken down into several smaller components, as indicated in Figure 2-26, forming the following links:

- *I/O Pad to V_{cc} (or V_{ss})*: For the connection between pad 1 and V_{ccp}, a simple diode can be used if the voltage of V_{ccp1} is higher than the pad 1

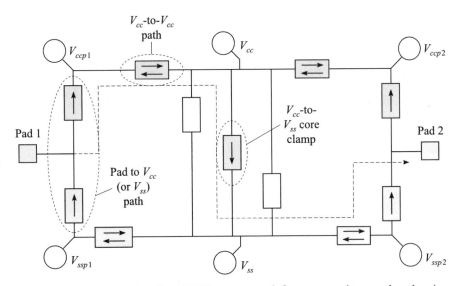

Figure 2-26. Example of preferred ESD current path from one pad to another showing the passage through I/O circuits and power supply clamps. Arrows indicating strongly directional devices are shown. It is also clear how the current path could terminate at any one of the power supply lines by passing through the elements shown.

voltage. If the pad 1 voltage is higher than V_{ccp}, then a suitable diode chain can be selected.

- V_{ccp} to V_{cc}: The connection between V_{ccp} and V_{cc} can also be established by a diode or a diode chain depending on the voltage difference between V_{ccp1} and V_{cc}. Also, the noise isolation between the two supplies should be considered.

- V_{cc} to V_{ss}: The path between the V_{cc} supply and V_{ss} is provided by the core clamp. This device is normally off during circuit operation. During an ESD event it is a low-impedance path and ideally a short.

- V_{ss} to Pad: The path between V_{ss} to pad is provided by the parasitic and a designed diode. Here one diode is sufficient and is also allowed.

Each of the components described above will be considered next. The path is more easily constructed and understood when all the elements are operating in the non-breakdown mode and they are simulatable. However, sometimes it is not possible to have such nicely behaved devices but the current path can still be used, only it is more difficult to have confidence in its as simplifying assumptions will have to be made. The advantages of a predictable path using simulatable devices over nonsimulatable devices are listed in Table 2-4.

As noted earlier, ESD current paths have been considered for some time [Merrill, 1993; Dabral, 1993, 1994; Maloney, 1995; Kever, 1997]. A recently implemented version of a current path scheme may look as in Figure 2-27. Here the diodes provide the links from pad to power supply and an NMOS core clamp between V_{cc} and V_{ss}. Using such a simple scheme, successful ESD protection has been reported [Merrill, 1993]. It is noteworthy that most authors reported success when the current path model was used. New work is being published on using densely-packed PMOS devices for power supply clamping, including stacked devices for high-voltage power supply clamping [Maloney, 1998].

Having discussed the practical benefits of a current path, the individual components will be discussed using a starting point the theory introduced in Section 1.4.

TABLE 2-4. Advantages of Simulatable over Nonsimulatable Devices

Advantages	Disadvantages
Simulatable	Not minimum area, resulting in larger capacitances
May not need upgrades	Needs constant monitoring with each technology variation (Spehar, 1994]
Minimizes experiments, faster turn-around time	Longer experiments and constant verification

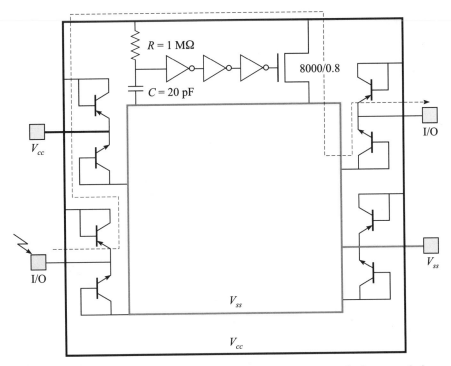

Figure 2-27. An ESD clamp implementation using the current path theory and simulatable NMOS shunt element [Merrill, 1993].

2.5.1. I/O Pad Segment

The pad-to-power-supply segment can have several options based on diodes, bipolars (if BiCMOS) or MOS. Of course, SCR-type options also exist, but these will not be discussed here.

2.5.1.1. Dual-Diode Based One approach to protect the pad is to use dual diodes. This is recommended primarily to save design and iteration time. It may be possible in some technologies to design better (more area efficient) clamps at the pad, rather than dual diodes, but these are usually very specific solutions and will probably need redesign when mapping to a new technology. One example is the use of BJTs in BiCMOS technologies.

On the other hand, diodes enjoy a favorable position when it comes to future technologies. First, the P^+-to-N^+ spacing decreases, which will also decrease the resistance directly. Also, as the N-well doping increases, the diode performance improves due to two factors: the resistance of the diode, which decreases [Voldman, 1994a], and current distribution, which is more uniform (i.e., has better diode area utilization). In one reported case, scaling the N-well resistance from 800 to 600 Ω/sq improved the HBM ESD performance from 4.7 to 6.4 kV [Voldman, 1994a].

When protecting the pad, the output NMOS is the weakest link. Usually, if the NMOS can be protected, then other devices will also be safe. Figure 2-28 shows a simple scheme to protect one such output buffer. In this method, a pair of diodes are added in parallel with the output transistors. This scheme works well in nonsilicided technologies. However, when silicided technologies are used, the performance drops dramatically [Amerasekera, 1995, p. 68]. There are two main causes:

- the nonuniform turn-on and snapback of the fingers and
- the loss of ballasting action that was present in the drain of the nonsilicided technologies.

It is important for silicided technologies to provide a method to distribute the current evenly over all the NMOS fingers. This may be done using an external ballast consisting of resistors or other techniques such as GCNMOS [Amerasekera, 1995, p. 69]. The objective is to encourage uniform current distribution in all the fingers.

If the ballast resistor option is chosen, the resistors can be implemented in polysilicon, diffusion, or N-well. The polyresistors have poor current-handling capability and too low a resistivity ($500\,\text{m}\Omega/\square$) to provide sufficient resistance and be area efficient. The diffusion resistors have been silicided so a special silicide-blocking mass has to be employed [Krakauer, 1992; Beebe, 1996], thus making the fabrication more involved. The only real option is the N-well

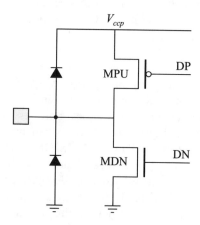

Figure 2-28. Simple dual diode protecting output buffers. This works well with the nonsilicided technologies.

resistor [Carbajal, 1992; Tong, 1996], which is implementable in a normal N-well P-epi CMOS process.

An implementation using N-well resistors is shown in Figure 2-29. Here specially placed N-well resistors are used in between the pad and the NMOS transistors [Carbajal, 1992]. The dual diodes are still used as the main current paths and the N-well resistor helps to limit the current flowing in each leg of the NMOS device.

It should be noted that the N-well resistance can be adjusted by varying the length (L_{nw}) and width (W_{nw}) of the N-well:

$$R_{nwell} = \frac{\rho}{t} \times \frac{L_{nw}}{W_{nw}} \tag{2-15}$$

where ρ/t is the N-well sheet resistivity. This relationship is true as long as the electric fields are low ($\ll 10^4$ V/cm) and the carrier velocity (V_d) scales linearly as Eq. 2-16; the current density is then related as

$$V_d = \mu E \tag{2-16}$$

and

$$J = q\mu_n E \times N_{nw} \tag{2-17}$$

where μ is the mobility and E the electric field. However, at high electric fields ($> \sim 10^4$ V/cm) the carrier mobility saturates at $\sim 10^7$ cm/s and the current

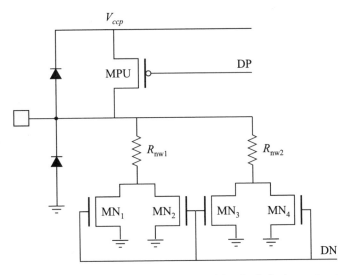

Figure 2-29. Output CMOS output driver protected by dual diodes and using N-well ballast resistors for the NMOS device.

saturates as shown in Eq. 2-18. This increase the effective N-well resistance dramatically:

$$J = q \times v_d \times N_{nw} \qquad (2\text{-}18)$$

This is valid if the N-well is significantly doped over the intrinsic silicon ($> 10^{14}\,\mathrm{cm}^{-3}$) but it is not degenerate ($> 10^{20}\,\mathrm{cm}^{-3}$). typical N-well dopings of $\sim 10^{17}\,\mathrm{cm}^{-3}$ gives V_{sat} of $10^5\,\mathrm{A/cm}^2$, or $10\,\mathrm{mA}$ in a 20-µm resistor [Antinone, 1986; Amerasekera, 1995]. Thus, it acts as a natural current-limiting element, and the current can be selected by controlling the width of the resistor. However, for the diode, the length should be scaled down to the least possible value. This allows a low-impedance path. Today it is possible to obtain diodes with a resistance of $\sim 2\,\Omega$. With scaling down of feature size permitting closer spacing of the P^+ and the N^+ and increase in N-well dopings, one should expect low resistance.

In Figure 2-30, the current characteristics of a diode and N-well are compared. The diode current is primarily limited by the *IR* drop, whereas

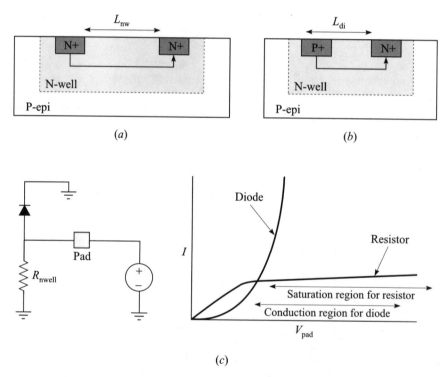

Figure 2-30. Comparison of (*a*) an N-well resistor cross section and (*b*) a diode cross section. Comparing the diode and the N-well current (*c*), it should be noted that whereas the diode current increases, limited primarily by the *IR* drop, the N-well resistor current saturates limited by the velocity saturation effect.

the N-well resistor current is limited by the velocity saturation effect. This is a simplistic model, but the N-well resistor assumptions have been validated by Amerasekara [1995, p. 162].

The above resistor and diode models are simplistic in the sense that they ignore the conductivity modulation that will occur in both elements. N-well conductivity increases as the minority carrier concentration increases above the background concentration. This transition occurs when the minority carrier (p) is equal to the background concentration. For 10^{17} cm^{-3} N-well doping (N_{nw}), the transition occurs at ~ 0.8 V, as given by the equation

$$V_0 = \frac{2kt}{q} \ln\left(\frac{2N_{nw}}{n_i}\right) \tag{2-19}$$

Since the conductivity is modulated, the resistance drops dramatically, enabling more current flow until the point when the diode is destroyed due to thermal heating.

Now consider the N-well resistor. The current through the N-well starts saturating when the electric field approaches 1×10^4 V/cm. Beyond this velocity in the saturation electric field and for more current to flow, the N-well carrier concentration has to be modulated. The voltage and current relationship is now given by [Antinone, 1986]

$$V = \frac{IL^2}{2\varepsilon_{Si}\varepsilon_0 V_{SL}} + (1 \times 10^4)L \tag{2-20}$$

and by differentiating the above, the space charge resistance can be found to be

$$R_{sc} = \frac{L^2}{2A\varepsilon_{Si}\varepsilon_0 V_{SL}} \tag{2-21}$$

The space-charge-limit resistance is plotted in Figure 2-31. The graph shows that by choosing a small diode length and a larger resistor length, the ratio of current flow in the diode to that in the resistor can be designed to be very large (> 10). For example, the ratio for a 0.4 μm-long diode to a 2-μm-long resistor can be ~ 30 (20,000 Ω/750 Ω). In this approximation both the diode and N-well were assumed to be conductivity modulated.

Next, if an NMOS device was added to the N-well resistor and the NMOS snapped back, the maximum current through one such NMOS device would be limited by the N-well resistor current. Now if the second breakdown current I_{t2} of a single finger in the NMOS was greater than the I_{sat} of the resistor, then the NMOS would be fully protected even if only one finger snapped back. This relationship is shown in the equation

$$W_{nwell} \times I_{sat} \leq W_n \times I_{t2norm} \tag{2-22}$$

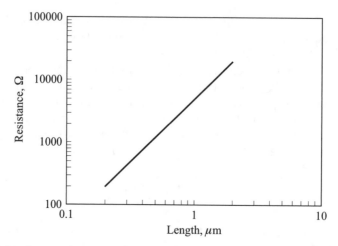

Figure 2-31. Saturated resistance of an N-well (1 µm deep, 10 µm wide) as a function of well length.

where I_{sat} is the resistor saturated current, W_n is the minimum NMOS single-finger width, W_{nwell} is the width of the N-well, and I_{t2norm} is the normalized second breakdown current per micrometer. A design using this approach would be as follows:

- Calculate the current drive or the impedance required for the I/O buffer.
- Use reliability data (or estimates) to calculate the minimum N-well width to carry the driver current or meet the impedance requirement.
- Use Eq. 2-22 to calculate the minimum width of the NMOS device from an ESD perspective.
- Scale the NMOS length to match the current and impedance requirements of the NMOS and N-well resistor when in normal circuit operation. If the minimum width required by the ESD requirements is smaller than the I/O width required, then the NMOS width can be increased to the required value. If the NMOS is too large compared to that required by the actual I/O driver requirements, then some legs can be tied to a normal off state.
- Section the resistor and NMOS into an appropriate number of segments to suit the layout.

Of course, using the GCNMOS or the ratioed-gate NMOS (RGNMOS) technique, all the NMOS fingers could be snapped back and ensure better uniformity in ESD current sharing.

Another method to increase the ratio of impedances between the NMOS path and the diode path is to utilize a pinch-type resistor in the N-well [Ghandhi, 1983; Orchard-Webb, 1991]. This is shown in Figure 2-32. The

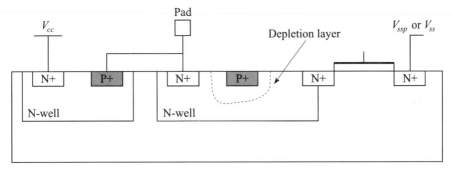

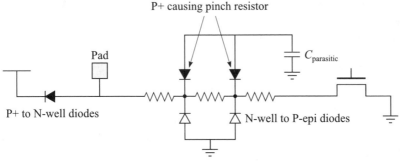

Figure 2-32. The ESD current through the resistor can be reduced by making a pinch resistor. Here the P$^+$- to N$^-$-well depletion layer and the N-well to P-epi (minor) modulate the cross-sectional area available for the current to flow. As shown, the P$^+$ charge shares between the N-well to P$^+$ and P$^+$ to others (parasitic). Since the parasitic capacitance is small, the P$^+$ sees the same voltage as the N-well. This will limit the effectiveness of the pinch resistors during the ESD event.

pinch resistor works on reducing the area available for the ESD current flow. As shown in Figure 2-32, the P$^+$ diffusion itself reduces the N-well area. Further, as the N-well is biased positive and (if) the P$^+$ is held at a lower voltage, then the depletion layer grows around the P$^+$ region. This pinches the N-well further. This action is similar to that found in a junction FET (JFET). However, once the voltage exceeds the breakdown voltage of the P$^+$ to the N-well, the pinching action due to the P$^+$ is reduced. However, the N-well to P-epi depletion layer can still grow. In some examples, the pinch resistor can approach an open circuit at high voltages (during ESD stress) and appropriate junction depths and concentrations. It should also be noted that the implanted N-wells have a retrograded doping profile. They are lightly doped on the surface and the doping peaks under the surface. In such N-well resistors, the current flow at small conduction levels is through this heavily doped region under the surface [Krabbenborg, 1991]. Therefore, at small pad voltages, the depletion layer is small, and this subsurface current path is not heavily affected. When the device is in normal operation, pad voltages are indeed smaller, especially when the NMOS is pulling down and the major portion of the N-well impedance is not hampered.

Next a special implementation of the pad protection is examined in BiCMOS technology. It should be noted that BiCMOS technology may have restricted use as the VLSI voltages scale down. This is so because the base–emitter voltage required to turn on a BJT becomes a significant portion of the supply voltage (e.g., V_{be} of 0.7 V is 28% of a 2.5-V technology and 39% of a 1.8-V technology). So the following scheme is not intended to be a mainstream methodology.

2.5.1.2. BiCMOS BJT Based
If BiCMOS technology is available, it is more area efficient to use the BJT to provide the main ESD clamping and to use the MOS for sensing and control functions. Such schemes have been developed elsewhere [Chatterjee, 1991; Mack 1992; Corsi, 1993].

A circuit utilizing this concept is shown in Figure 2-33 using computer-aided design (CAD) tools such as Pisces and SPICE. The circuit utilizes the large dv/dt of an ESD pulse to provide a trigger. The capacitor C_1 (parasitic) couples the charge to node N_2, which is discharged by R. This RC circuit acts as a high-pass filter, and for fast dv/dt it allows enough voltage buildup at N_2 to turn on MP_1 which in turn forward biases Q_1. The sizing of the resistor R_b is crucial. There is an upper limit on R_b. It cannot be too large as it is needed to turn off Q_1 during normal circuit operation and a large value would result in slow turn-off. Simultaneously R_b cannot be too small as then MP_1 would never be able to turn on Q_1 during and ESD zap. In this particular case an R_b value of 1 kΩ was selected [Chatterjee, 1991].

This method has two advantages:

- shorter ESD zap discharge path, thereby improving the robustness of the circuit, and

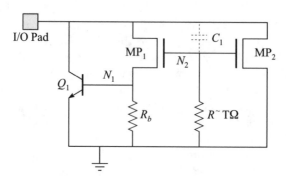

Figure 2-33. A BiCMOS clamp designed using circuit and device design tools.

- utilization of the I/O driver BJT as a clamp reducing the capacitive loading on the pad.

Using this circuit, ESD robustness gains of $11\times$ from 500 to 5500 V have been observed.

As mentioned earlier, the previous circuit may have capacitive coupling of noise or fast transients into the driver gate. The coupled noise can trigger the Q_1 into conduction. A trick that can be used to isolate the trigger is to employ MOSFET control on the BJT [Tandan, 1994]. This is shown in Figure 2-34. Here, MP_1 is used as a control switch to allow resistive coupling into the Q_1 base. When the chip is not powered up, V_{cc} is low. If now the pad is zapped positively with respect to V_{ss}, the MN_1 and MP_1 conduct and couple charge into the base of Q_1. This coupling allows Q_1 to conduct and dissipate the ESD stress. During normal circuit operation V_{cc} is high and MP_1 is turned off. Thus little coupling between the pad and base of Q_1 exists.

It should be noted that the 10-pF capacitor shown in Figure 2-34 is a fairly large capacitor for every pad. Current technologies (0.5 µm) have gate capacitance of ~ 3–$5 \, fF/\mu m^2$. So a 10-pF capacitor needs $\sim 3333 \, \mu m^2$ ($100 \times 33 \, \mu m$) of area. Repeating this area for every pad would be wasteful. It is prudent to have the sense signal generated once or in a few places on the die and then distributed to the various buffers.

This circuit can still turn on if large I/O transients that are not clamped exist on the pad. If the transient is much larger ($> V_{cc} + V_{tp} + V_{tn}$) than V_{cc}, MP_1 will turn on, and Q_1 will begin clamping the pad node. This clamping may or may not be beneficial in reducing overshoots. The rate of turn-on of Q_1 may

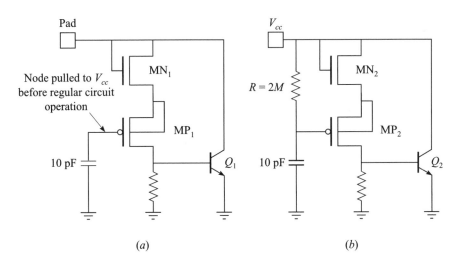

(a)   (b)

Figure 2-34. (a) A clamp based on BJT shunting for an I/O pad and (b) for V_{cc} protection.

cause very sharp transients that could be detrimental to the signaling scheme. This can be overcome by raising the trigger voltage of the clamp device. One such scheme that will achieve this is shown in Figure 2-35.

The clamp with the higher trigger voltage is based on using a circuit that is "on" when the chip is powered down [Tandan, 1994]. The 10-pF capacitor is chosen to keep the time constant ~20 µs far larger than an HBM ESD pulse duration. The MN_1 and MP_1 transistors are on initially when the ESD pulse strikes the pad. This enables a base current to flow into Q_1. The Q_1 is the main ESD shunt path. The PMOS MP_1 has to be turned off during regular circuit operations.

Having looked at some BJT and diode-based clamps, MOSFET options are examined.

2.5.1.3. MOS Based
These are truly not non-breakdown clamps as they will usually snap back when stressed. Some schemes rely on an ESD path from pad to V_{ss} provided by the NMOS device. The design aspect is to ensure uniform breakdown of all NMOS devices maximizing their robustness. These utilize schemes such as the GCNMOS and ballast techniques. Also, TFO types of MOS can be used if they are well ballasted.

Additional parasitic paths can be added, such as an NPN between the pad and the power supply for an ESD path. One such example is the addition of an N^+ tied to the I/O pad close to an N-well tied to the peripheral V_{cc} or core V_{cc} [Voldman, 1994a, c]. Since the spacing of this NPN can be varied to the desired

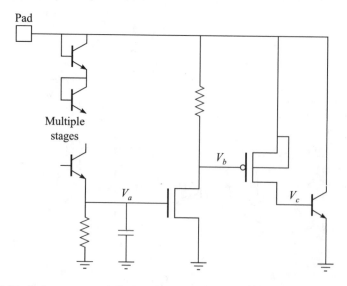

Figure 2-35. Series-connected diode chain can provide good detection capability for the ESD zap and provide sufficient isolation during I/O transients.

snapback point and there is an N-well ballast resistance in the path, the NPN snapback does not destroy the parasitic path.

Previously the pad-to-power-supply protection path was discussed. Next we will consider power-supply-to-power-supply clamping. Again these can be diode, BJT, or MOS based.

2.6. POWER SUPPLY COUPLING SEGMENT

A power supply coupling segment provides a path between two power supplies. Typically the two voltage levels are the same (or similar, e.g., V_{ccp}, V_{cc}, or V_{ddq}) only differentiated into a "clean core" V_{cc} and a "dirty" I/O V_{cc} (and V_{ss}). The differences in V_{ccp} and V_{cc} and V_{ssp} and V_{ss} will also be discussed. Unlike the pad protection circuit that must be present for each pad, these clamps are shared over a number of buffers. They can be diodes or MOSFET based. The diodes are easy to implement and area efficient. However, they can cross couple these supplies if there are large AC transients.

2.6.1. Diode Based

Diodes are effective cross-coupling clamps. The following factors need to be considered when constructing a clamp:

- maximum voltage difference between the supplies being clamped,
- noise margin desired during circuit operation before the diodes cross couple transients,
- maximum temperature of operation; and
- burn-in.

These factors determine the number of diodes used in the clamp between the noisy (V_{ccp}) and quiet (V_{cc}) supplies. If the diodes are ideal, increasing the number of diodes improves the cut-in voltage linearly. Cascading diodes also increase the diode resistance linearly. If the diode chain resistance is to be kept constant when diodes are cascaded, the diode area must be scaled up by a factor equal to the number of diodes (n) in the chain. Therefore, the total diode chain area scales up as n^2.

In a typical P-well CMOS process, however, the diodes are really PNP transistors, as discussed in Section 2.4.2.2. Increasing the number of diodes does not increase the cut-in voltage linearly and further additions show a diminishing return. The positive aspect of this PNP transistor action is that the resistance of the forward path is less than that expected in an ideal cascaded diode chain.

Once the diode turns on, it takes a low voltage to increase the current through them (or the AC resistance of a forward-bias diode is small). This is modeled in Figure 2-36. The larger voltage drop component then is due to the *IR* drop in the diode. Assuming there is uniform PNP transistor action in the chain, the input AC resistance can be calculated

$$R_n = R_d \left(1 + \frac{1}{(\beta + 1)^{n-1}} \right) \tag{2-23}$$

where R_d is the individual diode resistance and R_n the total resistance of the chain. For small β, the resistance of the diode chain decreases drastically. For very large β (at high currents) the resistance would approach that of a single diode since most of the current would be sunk to ground in the first stage. In comparison, for ideal diodes (no parasitics) the input resistance will be nR_d. In reality, the effective resistance is somewhat between the two extremes. When β is zero, the diode needs to be scaled in area as n^2 to maintain the chain resistance R_d. However, in a cascaded chain it is not necessary to scale the diode as n^2 but to a smaller degree, as shown by Eq. 2-23.

The V_{ccp}-to-V_{cc} clamping can be achieved by utilizing diode chains. If V_{ccp} and V_{cc} are of the same voltage magnitude, a symmetric and small chain can be used. However, if V_{ccp} and V_{cc} are different in magnitude, then, usually, an asymmetric diode chain results, as shown in Figure 2-37. In the symmetry aspect the V_{ssp}-to-V_{ss} clamp is different than the power supply clamps in two ways:

- There is a parasitic diode from V_{ssc} to V_{ssp} and no choice exists. From V_{ssp} to V_{ssc} the number of diodes can be again selected.
- The voltage difference between V_{ssp} and V_{ssc} is usually zero, and only noise coupling has to be considered.

As discussed earlier, the cascading diodes have two drawbacks: sublinear increase in forward voltage at a constant leakage current with diode additions to a chain and increased resistance in the ESD regime. This raises the question of whether it is possible to design the diode chain such that the following are true:

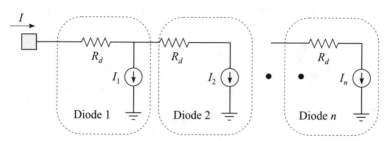

Figure 2-36. Observed resistance seen from the input decreases drastically when the PNP parasitic transistors cut in.

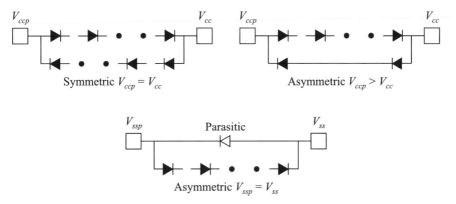

Figure 2-37. Diode-based coupling clamp on two similar voltage levels separated primarily because of noise issues. The DC voltage difference between V_{ccp} and V_{cc} can be large, whereas there is usually no voltage difference between V_{ssp} and V_{ss}.

- At ESD voltages the diodes utilize the PNP action to reduce the chain resistance.
- At low circuit operation voltages the diodes behave as linear diodes, as if no PNP transistor existed.

If this could be realized, then the larger cut-in voltage of the linear diodes during circuit operation and the Darlington action during ESD operation could both be utilized.

It is indeed possible to design such a circuit. The source of major leakages in the diode chain is the Darlington multiplication due to the cascading of the PNP transistors and the initiating leakage current at the end of the chain, which is magnified several times as it passes through the chain:

- To reduce the leakage, the Darlington multiplication can be reduced. This will make the diode chain behave more linearly and will have a corresponding penalty on ESD performance.
- The leakage current can also be reduced by reducing the number of stages through which the current must pass.

The second option is shown in Figure 2-38. Consider the leakage at the sixth diode (I_6). This has to come through all the transistors, and at each step it is amplified. If the entire I_6 leakage current can be supplied by a resistor, then no current needs to be supplied through the preceding Darlington stages, resulting in reduced overall leakage.

If this is true, and if it is assumed that the diode leakage is identical and the voltage drop across each resistor is identical, then the resistor values are given as

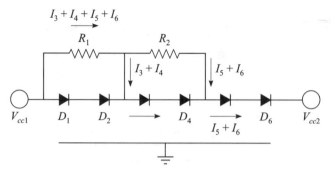

Figure 2-38. Bias network for a six-diode mixed power supply clamping string designed to maximize the temperature at which leakage current $\Delta V / R = 2.5 / R$ flows while using minimum total resistance.

$$\frac{R}{n-1}, \frac{R}{n-2}, \frac{R}{n-3}, \ldots, \frac{R}{3}, \frac{R}{2}, R \qquad (2\text{-}24)$$

In the resistor-biased case the last diode leakage has to be supplied by a resistor of value R. The next stage has to supply the last stage (I) and its own leakage current (I) at the same voltage drop. Therefore its resistance has to be halved. The next stage has to supply the previous two leakage currents and its own ($3I$); therefore, the resistance is $\frac{1}{3}R$. This repeats for each of the $n-1$ resistors required for an n-stage chain. This has indeed been observed (Figure 2-39).

This leakage is greatly exacerbated when the two different voltages (magnitudes) are being clamped. The detailed analysis and results are discussed in the mixed-voltage section in Chapter 7.

The diode option is area efficient, but it has issues due to the high leakage. It is possible to optimize the leakage with respect to the forward voltage of the

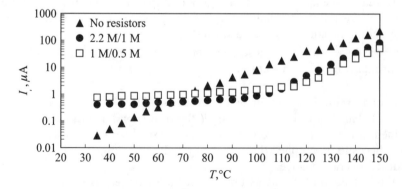

Figure 2-39. Measured leakage for a 3.0–5.5 V six-stage cladded diode string with discrete resistors as in Figure 2-38. Values were chosen to allow low leakage up to 100°C.

diode string using the technique developed by Maloney [1996]. The total voltage is maximized if, for a diode string n stages long, the total current I_{tot} is partitioned such that current I_j entering the jth diode satisfies $I_1 = I_2 = \cdots = I_{n-1} = (\beta + 1)I_{tot}/\beta n$ and $I_n = I_{tot}/n$. Circuits to reduce leakage have been developed and will be discussed in Section 2.7. Meanwhile the potential use of MOSFETs as power supply clamps is examined next.

2.6.2. MOS Based

Clamps that are MOS based have been utilized elsewhere [Murakami 1991; Dabral, 1994]. Although large, since they are repeated once for a number of buffers, they are still manageable. An important advantage of these clamps is that they can be turned off and no coupling exists during circuit operation. Simple versions employ either a diode-connected MOS [Murakami, 1991] or suitable trigger circuits [Dabral, 1994].

Here power-supply-to-power-supply clamps have been described. The idea was to reference all peripheral supplies to a common high-voltage node, V_{cc}. This approach has been employed by the authors and others [Kever, 1997; Voldman, 1994a] successfully. As an example, Kever [1997] referenced 19 independent power supplies to a common V_{cc} reference using this technique. To complete the path, the V_{cc} node must be clamped to V_{ss} and this is discussed next.

2.7. V_{cc}–TO–V_{ss} CORE CLAMPS

2.7.1. MOS Based

There are several options for designable core clamps depending on the technology and protection required. The simplest is to use an NMOS or PMOS [Maloney 1998] device to provide the clamping. An example is shown in Figure 2-27. The circuit consists of an RC sense element, a series of inverters for the correct logic and voltage drive, and the actual NMOS clamp. Such a clamp can be seen in Figure 2-40.

The NMOS clamp idea can be extended to I/O drivers as core clamps [Dabral, 1994]. Each CMOS I/O driver is a connection between V_{ccp} and V_{ssp} when both N and P devices are on. So, if during an ESD event we can arrange the predrivers such that they turn both devices on, a distributed clamp can be formed as shown in Figure 2-41. A rough estimate of the impedance of such a clamp consisting of 20 buffers each with 40 Ω on each side of the zap pad will be given as

$$R = \frac{R_{nch} + R_{pch}}{\text{number of buffers}} = \frac{40 + 40}{20} = 4\,\Omega \tag{2-25}$$

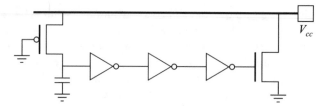

Figure 2-40. An NMOS clamp device and associated circuitry to sense and drive the clamp.

To couple the charge into the V_{ccp} and V_{ssp} line, diodes D_1 and D_2 can be used as shown in Figure 2-41. When the pad is zapped positive with respect to V_{ssp}, diode D_1 conducts charge to the V_{ccp} line. The powering of the V_{ccp} line enables the predrivers, which in turn switch on the NMOS and PMOS drivers. The ESD energy is used to power up the predrivers. It has been noted that such predrivers have ample time to turn on [Krakauer, 1994]. The voltage on the V_{ccp} line will build up to a point sufficient to dissipate the current. As the number of buffers employed in the clamp increases, the clamped voltage of the V_{ccp} should decrease, as shown in Figure 2-42. This is approximated by

$$V_2 = \frac{V_1}{\sqrt{N_2/N_1}} \tag{2-26}$$

where V_1 is the voltage with N_1 number of buffers and V_2 is the voltage with N_2 number of buffers.

There are a few issues with this clamp:

- It needs a number of buffers to implement.
- It needs the predriver to be modified.

Normally this modification is fairly simple but it will increase the size (small) as well as the predriver delay (small). The advantages are that a distributed clamp is available and no dedicated core clamp is required. The distributed clamp can reduce the interconnect voltage drop, thereby increasing the ESD robustness of the chip.

Two MOS-based clamps have been examined in this section. They can be separate explicit clamps, or I/O buffers can be ganged to form the clamp. Another option is to reutilize diodes and this is discussed next.

2.7.2. Diode Clamps

Voltages are scaling down in VLSI. This is in accordance with scaling rules that, in order to support higher frequencies, state that feature size must be decreased, which in turn entails lowering the voltage to maintain a constant

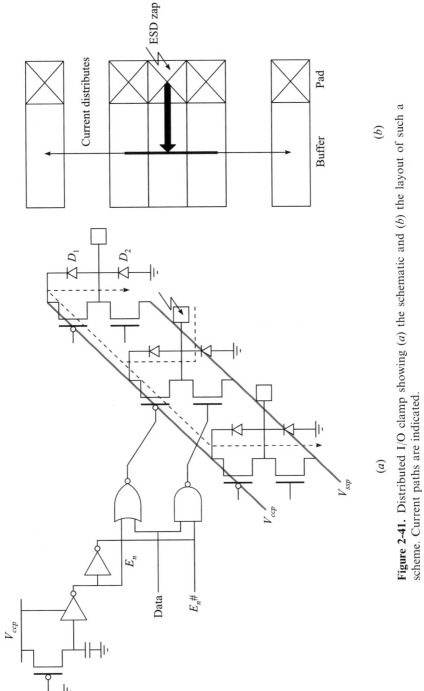

Figure 2-41. Distributed I/O clamp showing (*a*) the schematic and (*b*) the layout of such a scheme. Current paths are indicated.

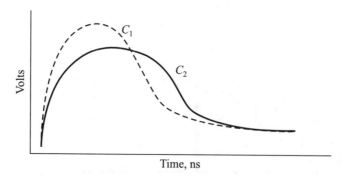

Figure 2-42. Voltage clamping due to I/O clamp as a function of the decoupling capacitor between the V_{ccp} and V_{ssp}. The greater the capacitor ($C_2 > C_1$), the lower the stress voltage. Similarly, the larger the number of buffers clamping, the lower the clamped voltage.

electric field. For these low voltages it is possible to simply use a small diode chain as a core clamp.

Simple experiments show that diodes acting as core clamps have some validity [Dabral, 1994]. Experimental results have indicated that even simple (unoptimized) chains of diodes can provide ESD protection. The circuit used in this experiment is shown in Figure 2-43. Three sets of circuits, each identical except for the core clamp, were constructed. For the diode chain, 14 diodes were stacked up to constitute a clamp. To compare diode clamp effectiveness, MSCR and LSCR were used as controls. These were drawn (modification of a standard SCR) by methods similar to those shown by Duvvury [1991]. The test results shown in Table 2-5 indicate that the first silicon diode chain clamp is on par with the SCR protection (which had been worked on for some time). Two points are worth mentioning:

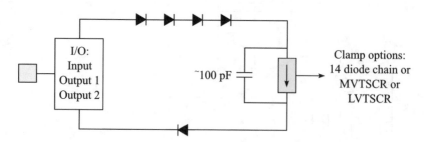

Figure 2-43. Experimental hook-up for core clamps to verify diode-based core clamp design.

TABLE 2-5. Diode-Based Core Clamp HBM and CDM Performance (kV) compared with SCR Protection

Core Clamp	CDM			HBM		
	Diode Chain	MSCR	LSCR	Diode Chain	MSCR	LSCR
Input	1.5	2	2	3	5	5
Output 1	1.5	0.5	2	3	4	3
Output 2	2	1	2	3	3	3

- As the capacitance ($\sim$100 pF in this circuit) increases, the core stress voltage will decrease and may delay SCR firing, whereas for the diode-based clamps, it will enhance performance.
- Other diode clamps (when placed in a real circuit) will assist in clamping, whereas that is not likely in an SCR clamp where one SCR will "hog" all the current.

In spite of the easy design of the diode clamp some concerns remain:

- The diodes may turn on at high temperatures and with noisy supplies during circuit operations. So it is essential that the forward voltage of the chain be increased above the worst-case maximum core voltages and noise estimates.
- The diode chain is designed to turn on above the worst-case operation voltage and temperature, which may be the burn-in voltage. Since these may be accelerated test conditions, larger leakage may be allowable but still they represent the harshest condition for the diode clamps. Consequently, the forward voltage is deliberately set high, leading to a large voltage drop on the ESD path. The larger forward voltage may degrade the ESD protection capability of the clamp.

Figure 2-44 shows a scheme that can improve the stand-off voltage and yet not be detrimental in the ESD regime. Here $\frac{1}{2}V_{cc}$ voltage is applied as an input to a unity voltage gain amplifier, which in turn clamps the midpoint to a $\frac{1}{2}V_{cc}$ voltage. This increases the cut-in voltage of the chain. This amplifier can only supply a current small enough for the leakage. As large current flows, such as during an ESD zap, the amplifier is effectively a high-impedance path, and its effect can be largely ignored. The diodes then behave as one large Darlington transistor [Maloney, 1994, 1995, 1996].

The simplest implementation of this circuit uses a source follower NMOS and is that shown in Figure 2-45. The reference voltage can be selected by adjusting the two PMOS devices T_1 and T_2. The voltage at the tapped node of the diode string is then one threshold voltage below the voltage ($V_{out} - V_{tn}$)

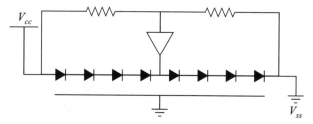

Figure 2-44. Use of a buffered voltage divider to supply extra bias current to the middle of a long diode chain. The leakage current to the last four diodes (near V_{ss}) is supplied by this amplifier, avoiding this leakage current from being multiplied by the last four stages.

set by the PMOS voltage divider. The PMOS voltage divider must be skewed if half of V_{cc} is to be obtained at the midpoint of the diode chain. So T_1 must be made wider and T_2 narrower.

The leakage current of one such boosted diode is shown in Figure 2-46. Such clamps were designed for a 2.5-V technology (maximum operating voltage of $2.5 + 10\% = 2.75$ V). Even with this simple implementation the leakage current is lowered by nearly 5–10 times over an unbiased case in the high-temperature range.

The boosted diode clamping performance was measured using a transmission line stress technique, and the results are shown in Figure 2-47. The eight-diode chain Boost8 turns on at 6 V but clamps well with an AC resistance of about 2.5 Ω (4 V, 1600 mA). Since this device will share current, any additional Boost8 clamps will lower this dynamic resistance linearly.

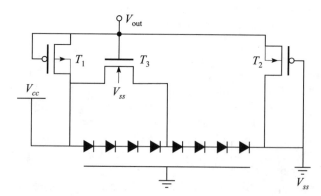

Figure 2-45. Boosted diode string as achieved with a buffered voltage divider in the form of a long-channel leader pair with T_1 and T_2, and a source follower T_3 supplying extra bias current to the middle of a long diode chain only when needed.

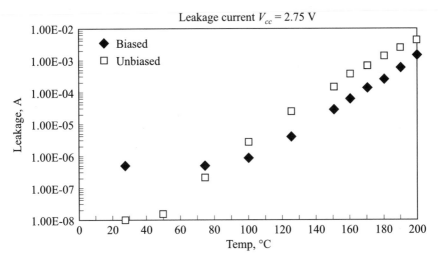

Figure 2-46. Current–temperature characteristics of an eight-stage diode string as drawn in Figure 2-45. The leakage current, with and without the booster network enabled are compared. This clamp is intended as a V_{cc} clamp for a low-voltage process.

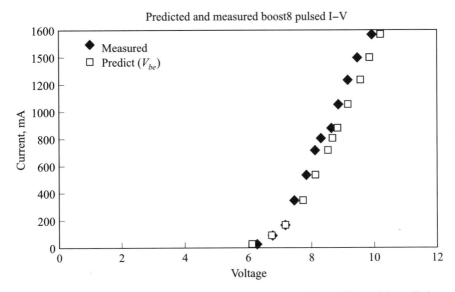

Figure 2-47. Boost8 pulsed $I–V$ measurements and comparison with model prediction.

2.7.3. Cantilevered Diode

The diode chain works well, but a serious drawback is that it can only turn on in excess of the operating or burn-in voltage (in addition, the noise margin must be built into the cut-in voltage). Thus there is a dead-band voltage when the clamps do not turn on until the stress voltage exceeds this cut-in voltage. Also the clamped voltage is greater than this cut-in voltage, and the leakage at the supply voltage can be substantial, especially at burn-in temperature. The question can then be asked if this dead-band voltage should be eliminated or diminished and can leakage be reduced.

It is possible to reduce this dead-band voltage by using a circuit "cantilevered diode" (Figure 2-48). It is so named because one of its terminals is hanging and the other is applied to the power supply [Maloney, 1995]. It is a one-terminal device.

To reduce the chain cut-in voltage, the number of diodes in a chain must be reduced. In the case of a boosted diode, a certain minimum number is required to support the V_{cc}-to-V_{ss} voltage. However, in the cantilevered diode case, the string length is independent of the requirement to support a voltage. Thus, by reducing the number of stages, reduced cut-in voltage should be expected, but the gain of the Darlington transistor will correspondingly suffer. The reduction in gain translates to a larger current appearing at the end of the chain that must be sunk by the switch.

In circuit implementation, the switches available are the PMOS, NMOS (CMOS), or BJT (BiCMOS). A PMOS terminated diode chain is shown in Figure 2-49. The timing is controlled by an RC timer set to $\sim 1\,\mu s$. Initially, when the V_{ccx} is low, the capacitor is in a discharged state. However, when the V_{ccx} is zapped, it takes some time for the capacitor voltage to build up, during which time the ESD pulse is dissipated. During normal circuit operation there is a $\sim 1\,\mu s$ start-up window where current from V_{ccx} will be sunk, the capacitor charges up, and the clamp turns off.

As noted earlier, the diode chain can leak at high voltage and temperature. So, biasing techniques similar to those used in the boosted diodes can be applied here, with one subtle difference: The end of the chain is terminated and not tapped into another power supply. The termination is disabled during normal operations. Thus the whole chain is at one potential and no linearizing of the voltage drop is necessary. If the Mpu transistors are made small, they will still function well to supply leakage and at the same time to reduce area.

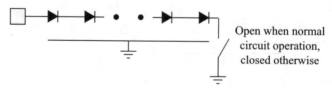

Open when normal circuit operation, closed otherwise

Figure 2-48. Simplified schematic of a "cantilevered diode" core clamp.

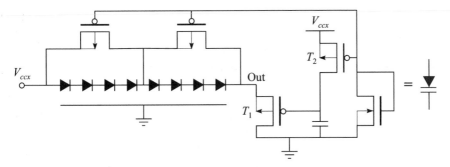

Figure 2-49. Six-stage cantilevered diode string with resistive bias network and termination circuit. The suggested circuit symbol could apply to any cantilever clamp network.

There will be no excessive leakage because during normal circuit operation the termination is turned off.

Figure 2-50 shows the results of one such six-diode-chain cantilevered diode. The cut-in voltage is much lower, ~3 V compared to 6 V for a boosted diode (Boost8), and the AC resistance is ~1 Ω (2.5 V, 2.5 A). The cut-in voltage is 3 V because all the diodes need to turn on. In part this can be attributed to the nonideal termination using a PMOS (ideally a switch), which needs at least a V_{tp} voltage at the out node to turn on. In addition, there is some charge

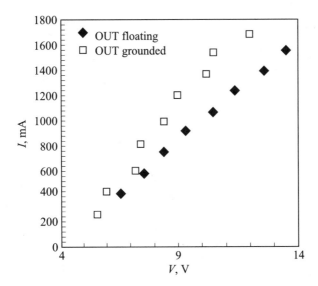

Figure 2-50. Pulsed I–V curves for six-diode cantilevered string in process P2 (see Table 2-3). OUT floating includes the required termination circuit and OUT grounded indicates how much voltage develops across that circuit.

coupled onto the capacitor due to the PMOS T_1 gate-to-source coupling, which raises the gate voltage, thereby raising the source voltage of T_1 before it can turn on. Due to the charge sharing, it is advisable to have a larger capacitance and a smaller resistance to obtain the desired RC time constant.

The cantilevered clamp can be modeled well using a simple Ebers–Moll model such as the one shown in Figure 2-23. The modeling was done on a spreadsheet with β, area and the current profile known, as shown in Figure 2-22 and Table 2-3. One such comparison between the model and experiment is presented in Figure 2-51. The model shows good correlation to about 1.5 A, after which there is departure. The model shows more conservative clamping capability at the high-current ranges. This ensures that the actual device will perform better than predicted, lowering the stress on the IC.

As the diode chain is shortened, the cut-in voltage diminishes but the current at the termination increases. Thus the PMOS T_1 will get larger to sink such current. Clearly a trade-off has to be made, based on β, the total clamp area, and the desired cut-in voltage.

The cantilevered clamps scale well with increased size, thus yielding a corresponding increased current capability. On a chip with several such clamps, they should all share in proportion to the voltage appearing at their terminals (subtracting IR losses of the interconnect leading to the clamp).

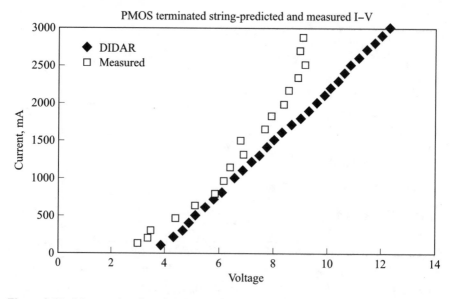

Figure 2-51. Measured and modeled pulsed I–V curves for a cantilever clamp in process P1.

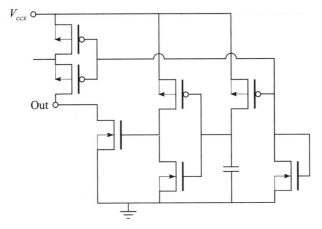

Figure 2-52. NMOS termination for a cantilever clamp. The last diode stage is connected to the OUT node. The diode bias network is also shown.

In the implementation discussed earlier the switch was a PMOS transistor, but an NMOS-terminated clamp can also be constructed, as shown in Figure 2-52. Here, a logical inversion is needed to power up the terminating NMOS, seen as an inverter inserted between the capacitor and the NMOS terminator. This inverter will cause a delay of ~ 100 ps (for $0.8\,\mu$m, lower for future technologies). The other aspects of the circuit remain unchanged. Clearly the NMOS clamp device is smaller because of the higher conductance of the NMOS.

In Sections 2.4 to 2.7, the clamps required to construct a current path were examined. These work well for the HBM pulse. However, because of their fast rise and fall times, CDM pulses create additional requirements. These issues will be discussed next.

2.8. CDM GUIDELINES

The previous sections have dealt with ESD phenomena targeted primarily at HBM pulses. For CDM protection, a few additional considerations are required. The CDM pulses are sharper and have larger current amplitudes so protection devices need to both be faster and have higher current capabilities.

In CDM testing four types of failure mechanism are observed [Maloney, 1988, 1992]:

- input gate destruction due to voltage buildup in metal buses due to the *IR* drop,

- a path to ground through the gate oxide for a substantial amount of charge leading to gate oxide destruction,
- an excessive gate voltage due to gate charging due to an avalanching junction leading to gate destruction, and
- metal burn-out due to high current density and bootstrapping action (this will be discussed in Chapter 5).

Each of these mechanisms is discussed next and solutions are recommended.

The source of all CDM failures is charge storage on the component. Components acquire charge due to triboelectric events, field-induced charging, and the like. When examining CDM-related damage and events, it is always useful to examine the package and chip layout for clues as to where the charge is stored, and how much total charge is likely to be involved.

The total charge of a CDM even on a packaged electronic component relates to its capacitance, which should scale roughly with the size of the component. The free-space capacitance of the component is the fundamental physical property involved. Because the free-space capacitance of a conducting sphere is approximately 1.1 pF/cm of the radius, each component can thus be considered to have an equivalent sphere, whose size relates to component size. The actual capacitance of a component in a real CDM event depends on ground plane position, of course, but free-space capacitance puts a lower limit on that value and sets the overall scale.

Exactly where the charge is stored is dictated by Gauss's law, which states that charge density on the surface of a conducting node is proportional to the electric field normal (i.e., perpendicular) to it. Thus, the bulk of the charge on a component is stored on the power planes, for example substrate V_{ss} and core power supply V_{cc}. The CDM damage almost always results when the charge on one or more of these power nodes becomes concentrated in a small location. Thus, when performing CDM failure analysis, the most important concept to consider is how and why one of these large chunks of charge found its way to the failure site.

Consider the input gate protection scheme shown in Figure 2-53. The current path shown is due to a negative CDM pulse, that is, the part that is charged negative with respect to ground. The grounded-gate clamp turns on, clamping the voltage across the gate. In the ideal case (a) where there is no bus resistance (or very little), even the passage of ~ 10 A does not cause the gate voltage of MN_1 and MP_1 to exceed the breakdown voltage. In reality there may be a substantial *IR* drop, in which case the voltage at the MN_1 and MP_1 gates will build up, which the grounded-gate device MGG cannot clamp due to the common mode voltage caused by the *IR* drop. This is shown in Figure 2-53b). If the input gate is removed from the immediate neighborhood of the clamp MTFO, then the MGG local clamp should be moved closer to the input gate or an additional one is added, as shown in case (c). This removes the

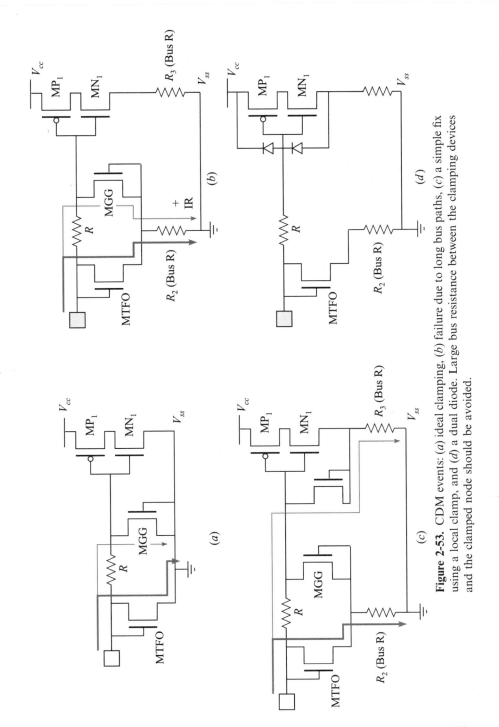

Figure 2-53. CDM events: (a) ideal clamping, (b) failure due to long bus paths, (c) a simple fix using a local clamp, and (d) a dual diode. Large bus resistance between the clamping devices and the clamped node should be avoided.

73

common mode *IR* drop for the clamps and MGG can protect the gate once more.

These solutions can be applied using diode clamps and are shown in Figure 2-53(*d*), and even the main clamp MTFO can be sometimes replaced by a dual diode. In the dual-diode case, voltage tolerance needs to be considered. When using the dual-diode scheme, the complete path to ground should be examined, as shown in earlier sections.

Charged-device model damage can also occur when the charge stored on the substrate, V_{ss}, is isolated with no nondestructive paths to the pad under the test. Consider the case in Figure 2-54. It shows the predriver, the driver, and a grounded-gate clamp at the pad. When there was no grounded-gate device, the ESD failure voltages were very low, while the grounded-gate device raised the ESD performance to acceptable levels. During a negative CDM event, damage to MPIO and MNIO drivers occurred. The damages are explained by the quick snapback of the predriver transistors MN_1 and MN_2 [Maloney, 1988]. The snapback time can be estimated as in the equation

$$t_{sb} \propto \frac{L_e^2}{2D} \tag{2-27}$$

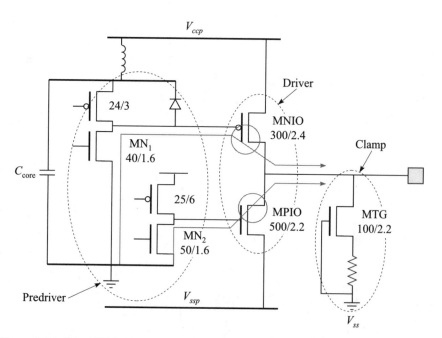

Figure 2-54. The CDM damage can occur in output devices MNIO and MPIO if the charge from the core finds the path through these devices easier than through other designed paths.

where t_{sb} is the snapback time, L_e is the effective channel length, and D is the carrier diffusion coefficient (in this case electron diffusion constant).

In the absence of MTG, the predriver transistors MN_1 and MN_2 devices (L is 1.8 and 1.6 μm, respectively) snap back fast, while MNIO, connected to V_{ssp}, cannot connect to the bulk of the charge on V_{ss}. This couples a significant portion of the substrate charge to the gate nodes of MNIO and MPIO, which breaks down the oxide, damaging it. The remainder of the charge finds it way through other parasitics (such as the P^+ N-well diode in the MPIO and the wire bond connections from V_{cco} to V_{cci} until the part is completely discharged.

Some relief comes from increasing the channel lengths of the predrivers, but with increasing predriver sizes comes larger propagation delays through the predriver and larger area and power dissipation. Thus, it may be preferable to use other clamps to achieve a harmless path to the pad for the charge stored on the substrate.

Another solution to prevent this oxide damage in the output transistors is to create alternative discharge paths where the major portion of the charge flows. These can be clamping diodes between similar supplies (V_{ccp} to V_{cc} and V_{ssp} to V_{ss}) [Maloney, 1990].

The previous example also indicates that it is preferable to tie the gates of the output driver using a "soft" supply instead of a hard-wired V_{cc} or V_{ss} connection [Lin, 1984, p. 208]. This is shown in Figure 2-55. the MOS devices

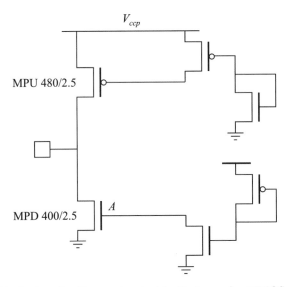

Figure 2-55. Method to tie off unused output buffer legs using NMOS pull-down and a PMOS pull-up. The pull-up and pull-down channel length should be such that they break down slower than the time it takes for the clamp to turn on. By having this tie-off arrangement, at least two gate oxides exist between the V_{cc} or V_{ss} power supply and the I/O pad, making it very robust.

used for tying off should have greater length than the grounded-gate device. For diode clamps, the tie-off devices should still be made larger than a minimum length to create a high-impedance path through the output device during a CDM event.

The input buffer can show stark dependence on timing of the snapback event. In Figure 2-56 a negative CDM zap is shown on an input pin. The damage is observed at the MN_1. The damage is ascribed to the snapback of the transistor MN_2 ($L = 2\,\mu m$) earlier than the MGG ($L = 5\,\mu m$). In this case, as soon as the MGG length was changed to a more appropriate 2µm, the CDM performance improved from well under 1500 V to over 1500 V. A dual diode can be considered to break the dependence of an input buffer to the grounded gate, as long as the voltage limits on the signals are acceptable.

The CDM pulses can cause damage in unlikely places. In the example shown in Figure 2-57, the gate oxide of a small pull-up device was damaged during a negative CDM test. The cause was traced to the passage of electrons to the gate, thereby charging it negative and increasing the gate drain bias at the PMOS and at the NMOS oxides. However, only the PMOS device breaks down. It is hypothesized that the conjunction of high voltages and the availability of charge carriers primarily cause the breakdown. Charges are more available where the field attracts majority carriers in the FET junction. This hypothesis when applied to the observed damage is shown in Figure 2-58. The PMOS junction has both the high field and the higher carrier concentration caused by the field attracting majority carriers into the P^+ junction and this is the reason for its breakdown [Maloney, 1988].

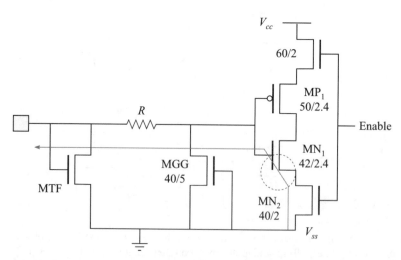

Figure 2-56. Damage to an input gate when the grounded-gate device is much slower than the input gates. When MGG was 40/5, the CDM performance was under 1500 V and when it was 40/2 it exceeded 1500 V.

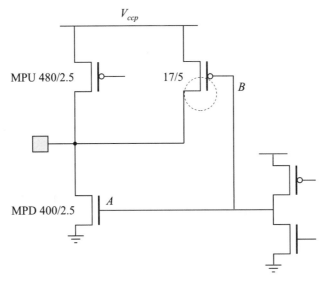

Figure 2-57. CDM damage due to the induced gate charging damaging the small pull-up device oxide.

The PMOS failure described earlier can be avoided by breaking the connection between nodes A and B. A separate predriver can then be placed for driving the PMOS and NMOS as shown in Figure 2-59. The negative-charge buildup at node A is not communicated to node B. This avoids the conjunction

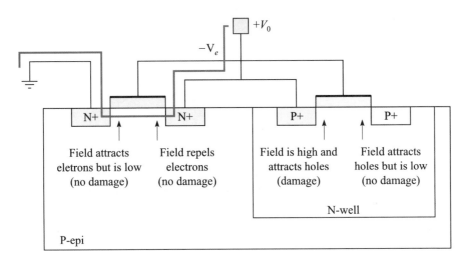

Figure 2-58. Gate-induced charge can be reduced by providing a path (diodes) or separating the predriver.

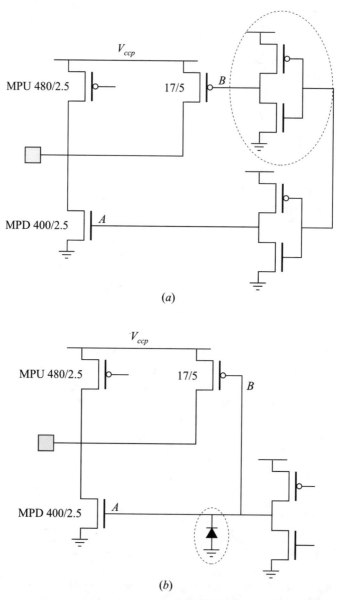

(a)

(b)

Figure 2-59. Two schemes to reduce the PMOS susceptibility to CDM pulses: (a) split the common node connecting the NMOS and PMOS or (b) add explicitly a P^+N^+ fast and low resistance diode. A P-epi to N^+ diode already exists in the drain diffusion of the NMOS tied to node A and may be satisfactory for ESD protection.

of high voltage and high carriers in the PMOS device. An alternative is to strongly couple the substrate to nodes A and B by a diode. This will prevent large negative voltage from building up at the PMOS gate, preventing damage.

To summarize, the protection of circuits from CDM and HBM stresses can use the same circuit techniques illustrated in this section. Usually CDM stress is more severe and faster so a design that is good for CDM will usually be good for HBM. The ESD pulse, once initiated on a pin, will affect the silicon. The main task of a design that mitigates the effects of this ESD pulse should be to maximize the ratio of current through designed ESD channels instead of the undesired paths. The paths at risk should have high impedance (if possible) and the desired paths should have low impedance. In this fashion ESD damage can be limited.

2.9. SUMMARY

In this chapter the following concepts were reviewed:

- Several case histories of ESD damage in "random"-path ESD protection schemes and their correction processes were presented.
- Two main classes of devices for ESD protection were identified as breakdown devices (TFOs, GGNMOSs, SCRs) and non-breakdown devices (MOSs, diodes, and BJTs).
- The disadvantage of the breakdown-type devices is that they are difficult to design, needing sophisticated models and physics. This process needs to be repeated every time a process parameter is significantly altered.
- The advantage of non-breakdown-type devices is that they are easier to simulate with a simple circuit simulator or a spreadsheet and are much less process sensitive. Further, their response can be simulated to verify operation in case any significant process parameter is changed.
- The "current" path and its elements were constructed using non-breakdown devices. Several types of clamps were examined: diodes, strings of diodes, "cantilevered" diodes, and I/O buffers used as clamps.
- The NMOS output transistor was identified as the most sensitive to ESD damage.
- A scheme to protect the output NMOS using an external ballast resistor was examined and a design method provided.
- The CDM design guidelines were elaborated. Prime among them are to reduce the bus resistance between the ESD clamp and the device to be protected and that the ESD clamp should be faster than the protected device.

• Especially for CDM, the driver, predrivers, and other associated circuitry on or near an I/O pad should break down much slower than the clamp protecting these circuits.

REFERENCES

[Amerasekera, 1993] A. Amerasekera, M. C. Chang, J. Seitchik, A. Chatterjee, K. Mayaram, and J. H. Chern, "Self-Heating Effects in Basic Semiconductor Structures," *IEEE Trans. Elec. Dev.*, **ED-40**. 1993, p. 1836.

[Amerasekera, 1995] A. Amerasekera and C. Duvvury, *ESD in Silicon Integrated Circuits*, Wiley, Chichester, England, 1995.

[Antinone, 1986] R. J. Antinone et al., *Electrical Overstress Protection for Electronic Devices*, Noyes, Park Ridge, NJ, 1986, p. 37.

[Baker, 1988] F. K. Baker and J. F. Pfiester, "The Influence of Tilted Source-Drain Implants on High-Field Effects in Submicron MOSFETs," *IEEE Trans. ED*, **35**(12), 1988, p. 2119.

[Bakoglu, 1990] H. B. Bakoglu, *Circuits, Interconnections and Packaging for VLSI*, Addison-Wesley, Reading, MA, 1990, p. 420.

[Beebe, 1996] S. G. Beebe, "Methodology for Layout Design and Optimization of ESD Protection Transistors," *Proc. EOS/ESD Symp.*, 1996, p. 265.

[Carbajal, 1992] B. G. Carbajal III, R. A. Cline, and B. H. Andresen, "A Successful HBM ESD Protection Circuit for Micron and Sub-Micron Level CMOS," *Proc. EOS/ESD Symp.*, 1992, p. 234.

[Chatterjee, 1991] A. Chatterjee, T. Polgreen, and A. Amerasekera, "Design and Simulation of a 4kV ESD Protection Circuit for a 0.5 µm BiCMOS Process," *IEDM*, 1991, p. 913.

[Corsi, 1993] M. Corsi, R. Nimmo, and F. Fattori, "ESD Protection of BiCMOS Integrated Circuits Which Need to Operate in the Harsh Environments of Automotive or Industrial," *Proc. EOS/ESD Symp.*, 1993, p. 209.

[Croft, 1994] G. D. Croft, "ESD Protection Using a Variable Voltage Supply Clamp," *Proc. EOS/ESD Symp.*, 1994, p. 135.

[Croft, 1996] G. D. Croft, "Transient Supply Clamp with a Variable RC Time Constant," *Proc. EOS/ESD Symp.*, 1996, p. 276.

[Dabral, 1993] S. Dabral, R. Aslett, and T. Maloney, "Designing On-Chip Power Supply Coupling Diodes for ESD Protection and Noise Immunity," *Proc. EOS/ESD Symp.*, 1993, p. 239.

[Dabral, 1994] S. Dabral, R. Aslett, and T. Maloney, "Core Clamps for Low Voltage Technologies," *Proc. EOS/ESD Symp.* 1994, p. 141.

[Darlington, 1953] S. Darlington, Semiconductor Signal Translating Device, U.S. Pat. 2,663,806, Dec. 22, 1953.

[Diaz, 1994] C. Diaz and G. Motley, "Bi-modal Triggering for LVSCR ESD Protection Devices," *Proc. EOS/ESD Symp.*, 1994, p. 106.

[Duvvury, 1988a] C. Duvvury, R. N. Rountree, and O. Adams, "Internal Chip ESD Phenomena beyond the Protection Circuit," *Proc. IEEE International Reliability Physics Symp.*, 1988, pp. 19–25.

[Duvvury, 1988b] C. Duvvury, R. N., Rountree, and G. Adams, "Internal Chip ESD Phenomena beyond the Protection Circuit," *IEEE Trans. ED*, **35**(12), 1988, p. 2133.

[Duvvury, 1991] C. Duvvury and R. Rountree, "A Synthesis of ESD Input Protection Scheme," *Proc. EOS/ESD Symp.*, 1991, p. 88.

[Duvvury, 1995] C. Duvvury and A. Amerasekera, "Advanced CMOS Protection Device Trigger Mechanisms During CDM," *Proc. EOS/ESD Symp.*, 1995, p. 162.

[ESDA, 1993] ESD Association Standard ESD Sensitivity Testing: Human Body Model (HBM)—Component Level, S5.1-1993.

[ESD, 1994] ESD Association Standard ESD Sensitivity Testing: Machine Model (MM)—Component Level, S5.2-1994.

[Ghandhi, 1968] S. K. Ghandhi, *The Theory and Practice of Microelectronics*, Wiley, New York, 1968, p. 324.

[Ghandhi, 1977] S. K. Ghandhi, *Semiconductor Power Devices*, Wiley, New York, 1977.

[Ghandhi, 1983] S. K. Ghandhi, *VLSI Fabrication Principles*, Wiley, New York, 1983, p. 630.

[Gray, 1993] P. Gray and R. Meyer, *Analysis and Design of Analog Integrated Circuits*, 3rd ed. Wiley, New York, 1993, p. 223.

[Horowitz, 1989] P. Horowitz and W. Hill, *The Art of Electronics*, 2nd ed., Cambridge University Press, 1989, p. 94.

[Hulett, 1981] T. V. Hulett, "On Chip Protection of High Density NMOS Devices," *Proc. EOS/ESD Symp.*, 1991, p. 90.

[Jaffe, 1990] M. Jaffe and P. E. Cottrell, "Electrostatic Discharge Protection in a 4-Mbit DRAM," *Proc. EOS/ESD Symp.*, 1990, p. 218.

[Johnson, 1993] C. C. Johnson, S. Qawami, and T. J. Maloney, "Two Unusual Failure Mechanisms on a Mature CMOS Process," *Proc. EOS/ESD Symp.*, 1993, p. 225.

[Ker, 1992] M. D. Ker, C. Y. Wu, and C. Y. Lee, "A Novel CMOS ESD/EOS Protection Circuit with Full SCR Structures," *Proc. EOS/ESD Symp.*, 1992, p. 258.

[Ker, 1997] M. D. Ker, C. Y. Wu, T. Cheng, and H. H. Chang, "Capacitor-Couple ESD Protection Circuit for Deep-Submicron Low-Voltage CMOS ASIC," *IEEE Trans. VLSI System.*, **4**(3), 1996, p. 307.

[Kever, 1997] W. Kever, S. Ziai, M. Hill, D. Weiss, and B. Stackhouse, "A 200 MHz RISC Microprocessor with 128Kb On-chip Cache," *Proc. ISSCC*, 1997, p. 410.

[Krabbenborg, 1991] B. Krabbenborg, R. Beltman, P. Wolbert, and T. Mounthaan, "Physics of Electro-Thermal Effects in ESD Protection Devices," *Proc. EOS/ESD Symp.*, 1991, p. 98.

[Krakauer, 1992] D. Krakauer and K. Mistry, "ESD Protection in a 3.3V Sub-Micron Silicided CMOS Technology," *Proc. EOS/ESD Symp.*, 1992, p. 250.

[Krakauer, 1994] D. Krakauer, K. Mistry, and H. Partovi, "Circuit Interactions During Electrostatic Discharge," *Proc. EOS/ESD Symp.*, 1994, p. 113.

[Kwon, 1995] K. Kwon, H. Park, D. Kim, K. Park, J. Jin, and S. Lim, "A Novel ESD Protection Technique for Submicron CMOS/BiCMOS Technologies," *Proc. EOS/ESD Symp.*, 1995, p. 21.

[Lin, 1984] C. M. Lin, L. Richardson, K. Chi, and R. Simcoe, "A CMOS ESD Input Protection Device, DIFIDW," *Proc. EOS/ESD Symp.*, 1984, p. 208.

[Lin, 1994] D. L. Lin and M. C. Jon, "Off-chip Protection: Shunting of ESD Current by Metal Fingers on Integrated Circuits and Printed Circuit Boards," *Proc. EOS/ESD Symp.*, 1994, p. 279.

[Mack, 1992] W. D. Mack and R. G. Meyer, "New ESD Protection Schemes for BiCMOS Process with Applications to Cellular Radio Designs," *ISSCC*, 1992, p. 2699.

[Maloney, 1985] T. J. Maloney and N. Khurana, "Transmission Line Pulsing Techniques for Circuit Modeling of ESD Phenomena," *Proc. EOS/ESD Symp.*, 1985, p. 49.

[Maloney, 1988] T. J. Maloney, "Designing MOS Inputs and Outputs to Avoid Oxide Failure in the Charged Device Model," *Proc. EOS/ESD Conf.*, 1988, p. 220.

[Maloney, 1990] T. J. Maloney, "Enhanced P$^+$ Substrate Tap Conductance in the Presence of NPN Snapback," *Proc. EOS/ESD Symp.*, 1990, p. 197.

[Maloney, 1994] T. J. Maloney, Electrostatic Discharge Protection Circuits Using Biased and Terminated PNP Transistor Chains, U.S. Pat 5,530,612, June 25, 1996.

[Maloney, 1995] T. Maloney and S. Dabral, "Novel Clamp Circuits for IC Power Supply Protection," *Proc. EOS/ESD Symp.*, 1995, p. 1.

[Maloney, 1996] T. J. Maloney and S. Dabral, "Novel Clamp Circuit for IC Power Supply Protection," *IEEE Trans. Comp. Pack and Manuf., Part C*, **19**(3), 1996, p. 150.

[Maloney, 1998] T. J. Maloney, "Designing Power Supply Clamps for Electrostatic Discharge Protection of Integrated Circuits," *Microelectronics Reliability*, to be published Nov. 1998.

[Mayaram, 1991] K. Mayaram, J. H. Cern, L. Arledge, and P. Yang, "Electrothermal Simulation Tools for Analysis and Design of ESD Protection Devices," *Tech. Digest IEDM*, 1991, p. 909.

[Merrill, 1993] R. Merrill and E. Issaw, "ESD Design Methodology," *Proc. EOS/ESD Symp.*, 1993, p. 233.

[Murakami, 1991] Y. Murakami, Integrated Circuit Apparatus Induced Static Electricity Protection Circuit, U.S. Pat. 5,034,845, July 23, 1991.

[Orchard-Webb, 1991] J. H. Orchard-Webb, "A Characterization of Components for an Optimized CMOS Input Protection System," *Proc. EOS/ESD Symp.*, 1991, p. 83.

[Pierret, 1990] R. F. Pierret, *Field Effect Devices*, 2nd ed., Addison-Wesley, Reading, MA, 1990.

[Polgreen, 1992] T. Polgreen and A. Chatterjee, "Improving the ESD Failure Threshold of Silicided *n*-MOS Output Transistors by Ensuring Uniform Current Flow," *IEEE Trans. RD.*, **39**(2), 1992, p. 379.

[Spehar, 1994] J. R. Spehar, R. A. Colclaser, and C. B. Fleddermann, "The Effect of Oxidation of the PolyGate on the ESD Performance of CMOS ICs," *Proc. EOS/ESD Symp.*, 1994, p. 257.

[Sze, 1981] S. M. Sze, *Physics of Semiconductor Devices*, 2nd ed., Wiley, New York, 1981.

[Tandan, 1994] N. Tandan, "ESD Trigger Circuit," *Proc. EOS/ESD Symp.*, 1994, p. 120.

[Thomas, 1979] G. B. Thomas and R. L. Finney, *Calculus and Analytic Geometry*, 5th ed. Addison-Wesley, Reading, MA, 1979, p. 620.

[Tong, 1996] N. Tong, R. Gauthier, and V. Gross, "Study of Gated PNP as an ESD Protection Device for Mixed-Voltage and Hot-Pluggable Circuit Applications," *Proc. EOS/ESD Symp.*, 1996, p. 280.

[Tsividis, 1980] Y. P. Tsividis, "Accurate Analysis of Temperature Effects in Ic-Vbe Characteristics with Application to Bandgap Reference Sources," *IEEE J. Solid State Circuits*, **SC-15**, 1980, p. 1076.

[Tsividis, 1987] Y. P. Tsividis, *Operation and Modeling of the MOS Transistor*, McGraw-Hill, New York, 1987.

[Voldman, 1994a] S. H. Voldman, "ESD Protection in a Mixed Voltage and Multi-Rail Disconnected Power Grid Environment in 0.50 and 0.25 μm Channel Length CMOS Technologies," *Proc. EOS/ESD Symp.*, 194, p. 125.

[Voldman, 1994b] S. Voldman and G. Gerosa, "Mixed-Voltage Interface ESD Protection Circuits for Advanced Microprocesors in Shallow Trench and LOCOS Isolation CMOS Technologies," *Proc. IEEE Int. Electron Devices Meeting*, 1994, p. 277.

[Voldman, 1994c] S. H. Voldman, S. S. Furkay, and J. R. Slinkman, "Three-Dimensional Transient Electrothermal Simulation of Electrostatic Discharge Protection Circuits," *Proc. EOS/ESD Symp.*, 1994, p. 246.

[Wallash, 1994] A. J. Wallash and T. H. Hughbanks, "Capacitive Coupling Effects in Spark-Gap Devices," *Proc. EOS/ESD Symp.*, 1994, p. 273.

[Webster, 1954] W. M. Webster, "On the Variation of Junction-Transistor Current Amplification Factor with Emitter Current," *Proc. IRE*, **42**, 1954, p. 914; quoted in S. M. Sze, *Physics of Semiconductor Devices*, 2nd ed. Wiley, New York, 1981, p. 142.

[Wheatley, 1976] C. Wheatley and W. Einthoven, "On the Proportioning of Chip area for Multistage Darlington Power Transistors," *IEEE Trans. Electron Devices*, **ED-23**, 1976, p. 870.

[Yang, 1978] E. S. Yang, *Fundamentals of Semiconductor Devices*, McGraw-Hill, New York, 1978.

CHAPTER 3

ADDITIONAL ESD CONSIDERATIONS

The primary focus of Chapter 2 was ESD current path implementation. Several additional factors that influence ESD reliability of a circuit and other design considerations will be discussed in this chapter. Primary among them are the effects of die capacitance, packaging options, antenna diode issues, hot-electron damage, and predriver design on ESD reliability.

3.1. CAPACITOR BENEFITS IN STRESS REDUCTION

Capacitors are implemented in VLSI circuits to reduce the on-chip power supply ripple. The larger the ripple, the larger the uncertainty in the chip timing, which in turn degrades the overall performance of the chip. The large di/dt required by the core switching cannot usually be supplied directly from the board or package because of the relatively large wire bond (or solder bump technology) inductances. Therefore on-chip capacitors are essential. This on-chip capacitance has two components, one that is explicitly added (gate capacitance) and the other due to parasitic capacitance of the core circuits (PMOS N-well to substrate, metal to metal). The other benefit of the core decoupling capacitor is to dramatically reduce the ESD stress between the V_{cc} and V_{ss} nodes due to charge sharing. The current die capacitances range from 1 nF for small dies to 100 nF for large dies. With future technologies having lower voltages and higher frequencies and currents, increasing die capacitance should be expected.

How much ESD stress is reduced by having a large on-chip capacitance? To help answer this question, consider Figure 3-1, which shows the HBM and

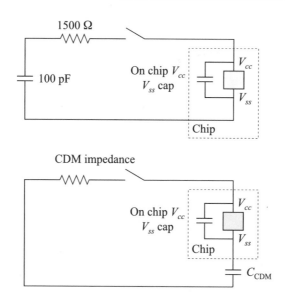

Figure 3-1. The HBM and CDM models stress the V_{cc} pins of a chip.

CDM models as they appear to the tester when an ESD zap is applied between V_{cc} and V_{ss}. The decoupling capacitor is directly in parallel with the ESD path and separated from the ESD capacitor with a considerable impedance; therefore, the decoupling capacitor does not allow the charge to build up immediately in the HBM case. In the CDM case the parasitic impedances are small; therefore, the decoupling and ESD capacitors share the stress voltage immediately. The effect of the capacitors is shown in Figure 3-2, where the maximum stress voltage experienced between V_{cc} and V_{ss} is plotted as a function of the decoupling capacitor. As the capacitance increases, the stress voltage experienced between the decoupled node decreases [Jaffe, 1990; Dabral, 1994; Ramaswamy, 1996]. The stress voltage may decrease to such a degree that many clamps with high trigger voltages may not function. An example is an SCR that may need a trigger voltage of 15–50 V. The trigger voltage may not be reached by the core supply (V_{cc} and V_{ss}) to trip on some clamps, but this voltage can be high enough to cause damage to the core circuitry. In general, clamps requiring high trigger voltages to initiate will be adversely affected by increasing decoupling capacitance.

The previous example dealt with a zap directly between V_{cc} and V_{ss}. When the ESD stress is between different I/O pads, the decoupling capacitor will still come into play. It effectively AC clamps the V_{cc} node to the V_{ss}, providing an easy path for AC discharging the V_{cc} mode.

Consider a positive CDM zap case. The equivalent model is shown in Figure 3-3 [Maloney, 1990]. The substrate is charged positive, and so all the nodes in the chip charge up positively. When the I/O pin is shorted, the V_{ss} (on-chip) to

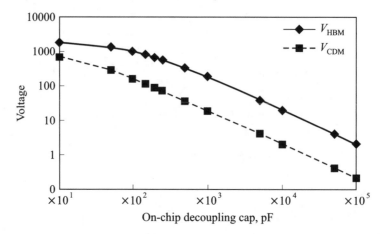

Figure 3-2. Charging device capacitance can drastically reduce ESD stress voltage experienced by the core circuits. This is particularly true for HBM-like voltages. For this analysis, the C_{CDM} capacitance has been assumed to be 20 pF and the ESD zap voltages for HBM and CDM are 2 and 1 kV, respectively.

ground capacitor discharges rapidly through the diode. Consider the case where the on-chip decoupling capacitor is small and there is a slow power supply clamp. For this case, the V_{ss} collapses rapidly, trapping charge Q in the V_{ccp} and V_{cc} nodes. This in turn stresses the I/O transistors MNIO and MPIO and the core transistors MP_1 and MN_1. In addition, the large $V_{cc}-V_{ss}$ buildup can cause snapbacks, latchup, and circuit destruction. If the decoupling capacitor is large, the same charge Q is trapped, the voltage developed across the V_{cc} and V_{ss} is much smaller and a slower buildup occurs. This provides a better chance for the clamps to function and safely remove the charge. Other nodes (V_{ccp}) coupled to the V_{cc} by the ESD diodes also take advantage of this large decoupling capacitor, and only their clamped voltage clamp V is higher than the V_{cc} node, as shown in Figure 3-3c.

The decoupling capacitors help the reduction in power supply ripple during circuit operation and also help reduce the ESD stress during an ESD zap. Large decoupling capacitors are therefore advisable.

3.2. PACKAGING EFFECTS ON ESD

3.2.1. Conventional Packaging

Packages can influence ESD performance significantly. The main considerations are:

- Overall package size (for CDM),
- substrate tap resistance,

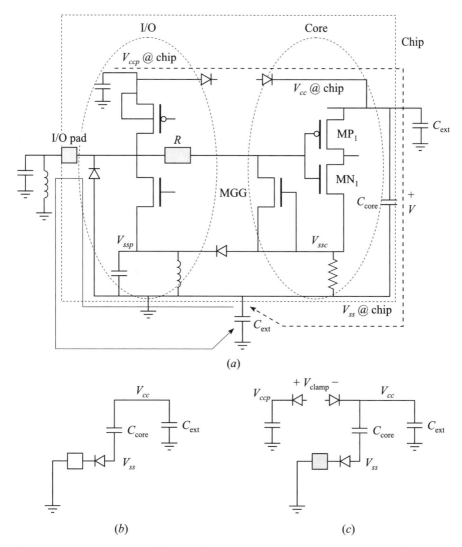

Figure 3-3. (*a*) Schematic of CMOS die representing CDM event at I/O pin. (*b*) Charge sharing between the V_{cc} node external capacitance (C_{ext}), core capacitance (C_{core}), and (*c*) clamped voltage of V_{ccp} node.

- decoupling capacitors on the package, and
- power supply partitioning.

Each will be examined in turn.

First, the overall package size is important for CDM because it determines the amount of charge that can be stored on the package. As stated in an earlier chapter, a package can be thought of as being equivalent to a sphere of some size, and a conducting sphere has an intrinsic capacitance in free space ($CV = Q$ with respect to ground at infinity) or about 1.1 pF/cm of radius. Thus the basic charge–voltage–energy relations are determined for a given package. In a CDM event, since all the charge on the package must be exhausted through a given package pin, a pin thus "feels" the total size of the package with larger packages generally causing harsher CDM events.

Now the effect of the substrate tap is considered. Some packaging options offer low-resistivity substrates. For example, in ceramic packages this is done using a gold eutectic bonding technique [Koopman, 1989; Shukla, 1985]. In plastic packages, an electrically conductive epoxy may be used. This bond is shown in Figure 3-4. Substrate contact can be effectively used to reduce the plane resistance during ESD (and circuit operation). Thus placing a good electrical contact between the chip and package is greatly beneficial. However, there are technologies that do not use this technique or have the option of which substrate to use as a power distribution, and on-chip metal has to be used to deliver power.

Another significant package-related effect that is relevant to CDM is the package capacitance to the charging plane and the package decoupling capacitors. for example, in CDM testing, the dielectric distance (T_d) between the external charge plate and the die and package defines the capacitance of the package. As the dielectric thickness increases, it forces the external charging plane away from the chip V_{ss} plane, resulting in lower capacitance. The reduced capacitance results in lower charging for the same voltage and the energy stored in the capacitor is lower. The reduced capacitance results in lowering the CDM current. In contrast, the HBM performance is not affected by the package dielectric thickness.

The package capacitance interaction is also determined by the type of pulse. For the HBM pulse the time constants are slow enough (~ 10 ns) that the ESD pulse can electrically interact with the package decoupling capacitance.

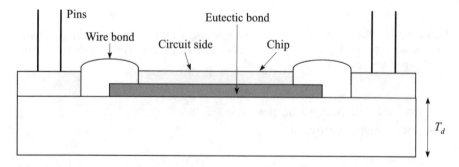

Figure 3-4. Regular PGA package showing he eutectic bonding used to attach die to the ceramic carrier. This eutectic bond provides a good electric short in the Si back, which affects electrical and ESD performance.

Normally, a chip will have a hierarchy of decoupling capacitors: on-chip, on-package, and so on. Thus the V_{cc} and V_{ss} capacitances on the package also have a chance to contribute to the ESD protection. The package capacitance for a large die may be easily in the 10–100-nF range. As seen earlier, this directly lowers the voltage stress on the V_{cc} and V_{ss} plane. For a CDM pulse the transient is much faster (~ 1 ns); therefore, the package capacitance has very little chance to interact with the die at the initial ESD zap time and relatively little benefit should be expected.

In the future, currents will increase from the present levels of the 10–20 A [Jain, 1997; Cohen, 1997]. The switching transients will also be fast, typically less than 0.5 ns. This leads to large di/dt (~ 20–40 A/ns). Simultaneously reduced power supply voltage is essential to maintain scaling. Thus, the tolerable voltage drop on the power supply (both DC and AC) is smaller. Therefore, for successful power delivery considerations, the high-performance package needs to have a very low inductance path and needs to couple the chip to the package very well. Now, the 10–20 A current and the transients of 0.5 ns and smaller are already comparable to the rise times of the CDM event. Thus, as the performance of the power delivery is improved, the CDM event should increasingly be able to interact with the package capacitance and derive benefit from that.

The package power options together with the on-chip power hook-up are another factor to consider for ESD reliability. Currently three common options are exercised in connecting power to the die:

- The clean (core V_{cc} and V_{ss}) and dirty (peripheral V_{ccp} and V_{ssp}) suppliers are not separated.
- The clean and dirty supplies are separated on the die but shorted on the package.
- The clean and dirty supplies are separated on the die and package and shorted only on the board.

These choices are based on switching transients, the amount of current switched, and the noise isolation desired. Typically, all three options have been used and are shown in Figure 3-5. If the power supplies are separated on the die, then an ESD path has to be created on the die, as shown in Figure 3-6. A simple solution is to use diode chains. Using MOSFETs has also been suggested but such coupling will be area consuming. Also a core clamp can be used to connect the power supply to ground, thereby avoiding a lengthy path.

When power planes are shorted on the package but not on the die, the ESD paths should be created in the die. The package parasitics may be high and may effectively act as high impedance for a CDM event, leaving the peripheral and core supplies unconnected. In this case, the nodes on the peripheral supply will find random paths to discharge and may lead to failure. Such failures have been reported by Wei [1993].

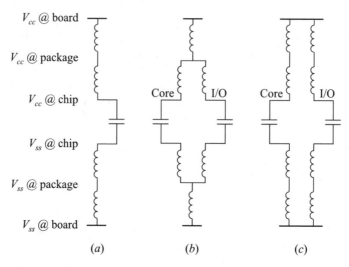

V_{cc} @ board

V_{cc} @ package

V_{cc} @ chip

V_{ss} @ chip

V_{ss} @ package

V_{ss} @ board

(a) (b) (c)

Figure 3-5. Three options of distributing power common in IC packaging: (a) peripheral and core supplies shorted on the chip; (b) peripheral and core supplies shorted at the package level; and (c) peripheral and core supplies shorted at the board.

The V_{ss} power plane choice has an impact on the ESD performance of the NMOS transistor, as illustrated in Figure 3-7. If the substrate and the NMOS source are not tightly connected using metal, then the only other connections are through the parasitic N^+ and P diode and through any additional clamps between V_{ssp} and V_{ss}. During NPN snapback, the substrate current has to return through the P^+ pickup, which ends up raising the substrate potential. This occurs until the forward biasing of the substrate-N^+ diode happens. With increasing substrate bias the snapback voltage V_{t1} decreases rapidly, whereas the second breakdown voltage V_{t2} remains approximately constant. This behavior enables all the NMOS transistor legs to snap back ($V_{t1} < V_{t2}$) before any go into a second breakdown. This in turn allows full utilization of the NMOS ESD capability. If the V_{ssp} and V_{ss} are tied well on the chip, then the modulation of V_{t1} does not occur and no corresponding benefits result [Polgreen, 1992].

3.2.2. Multichip Modules

Electrostatic discharge in multichip modules (MCMs) presents an interesting case in trade-offs in terms of area, ESD performance, and I/O speed. The aggressive I/O speed requires minimum ESD-related loading and design rules. Also, in an MCM, those pins that interact only with other pins within the MCM are never exposed to the external world after manufacturing (see Figure 3-8). Therefore, it is possible to scale back some ESD requirements, which in turn will reduce area. Although the I/O area in a typical chip may be

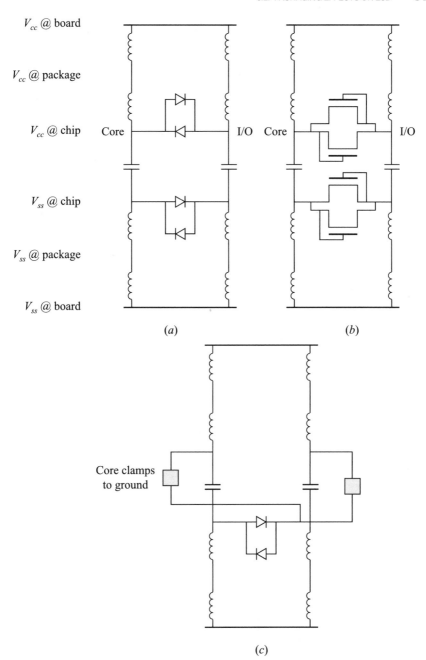

Figure 3-6. For isolated power supplies on a die, ESD paths should be created as shown in (*a*) using diodes, (*b*) using NMOS, and (*c*) using cantilevered clamps.

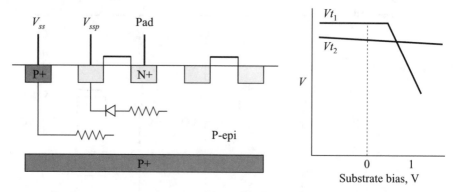

Figure 3-7. Maximum ESD capability of an NMOS is achieved if the peripheral and core V_{ss} buses are split [Polgreen, 1992].

$\sim 5\%$, it is still a significant portion of the silicon and important to save area. To place this area in perspective, current large dies may be $\sim 1.5 \times 1.5$ cm, making 5% ~ 0.1 cm^2 (0.075×0.075 cm). This size is comparable to a number of current smaller and older generation chips.

To enable area optimization on pins never exposed to the external world, ESD structures must be scaled down to meet the revised target specifications. What should be the ESD protection required for these structures? As a pessimistic guess, they can be designed as if they were normal I/O pins seeing the external world. However, in reality, these pads only see the external world during fabrication and assembly. Currently, one can only guess at this answer, as no solid data are available. In the future with increasing use of MCMs this may be an important consideration. If it is possible to scale back the ESD

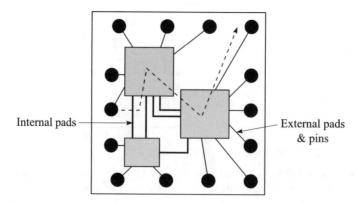

Figure 3-8. An MCM interconnection scheme showing internal pads and external pads and pins visible to outside world.

requirements, then the protection devices can be scaled back, thereby resulting in area reduction and reduced capacitive loading on the bus.

It is possible that in an MCM only one chip has external pins and all others do not see external pins. In such a case normal ESD protection relevant to the technology will suffice. However, it is possible that there may be multiple independent power supplies needed for the various chips. A good example may be a processor on one logic (N-well) technology, and the memory in a DRAM (P-well) technology, running on different voltages, as shown in Figure 3-9. Clearly, to be fully protected, a good ESD path needs to be created between the two separate power pins. This will clearly need a common path in the package and individual protection schemes in the two affected chips. In Figure 3-10 one such scheme is shown. Here the common path is the V_{ss}, which is shared by all the chips. It should be noted that in single-chip packages there is degradation of ESD robustness with longer signal traces and wire bonds [Voldman, 1992]. Thus great care should be taken to minimize the impedance in such paths (both DC resistance and AC transmission line behavior). Such connecting power planes should be wide.

3.3. SMALL-CHIP ESD ISSUES

In spite of the general trend of increasing die size and correspondingly larger core capacitance, many small chips are in existence and others will continue to be designed in the future. These may not have very large decoupling capacitors and may be limited to a nanofarad range (or less). The stress on V_{cc} and V_{ss} will correspondingly increase (~ 200 V for HBM and ~ 20 V for CDM from

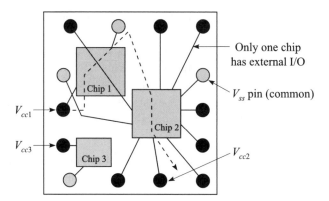

Figure 3-9. A module may have multiple independent power supplies supporting the various chips. An ESD zap path between one power pin to another power pin needs to be protected.

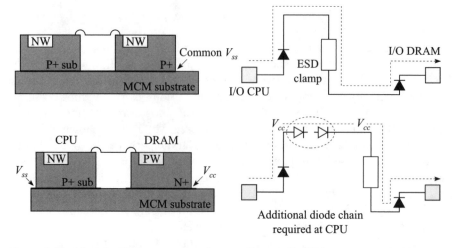

Figure 3-10. Module ESD protection is complicated if different types of chips are assembled onto an MCM substrate: (*a*) N-well CMOS Central processing unit (CPU) is placed with N-well DRAM; (*b*) N-well CMOS CPU is placed next to P-well CMOS DRAM. In the latter case a common substrate is not possible so either the V_{cc} or V_{ss} or both can be bridged.

Figure 3-2). So, an efficient and fast core clamp is essential. Additional design recommendations for these small chips are as follows:

- Increase the number of core clamps so AC resistance is low when the clamps turn on.
- Increase the die capacitance by deliberate decoupling capacitor placement; this is usually expensive but the circuits get to share the benefit.
- Increase the number of power supply coupling diodes to reduce their AC resistance.
- Increase the number of V_{cc} and V_{ss} bond wires to the package and also increase the package plane capacitance. This effectively allows the package capacitance to help reduce the on-chip stress.
- If a number of small power buses exist, reference them to V_{cc} or a similar high-capacitance node (as far as possible).

Unfortunately, the issue is compounded by small on-chip decoupling capacitors, a small number of power supply wire bonds, and maybe by low-cost packaging with no decoupling capacitor options, making these chips very hard to protect. If in addition these chips have to support multiple different voltages, as required by some interface chips, the ESD issue is further aggravated.

Usually the design of these chips is limited by tight economic pressures, and increasing the die area for decoupling capacitors and changes in packaging are strongly discouraged. Due to the technical and economic reasons compounded, *these will constitute a major ESD reliability threat in the future.*

In the last section, the bypass capacitor was discussed as one electrical element, but in reality it is distributed over the entire chip. By their additive property, the sum of these capacitors becomes a significant contributor to ESD protection. The same can be said for active clamps (as well as for the NMOS device where deliberate efforts are made for them to share the ESD stress evenly).

3.4. BENEFITS OF DISTRIBUTED CLAMPS

The benefits of core clamps have been discussed in the previous chapter. For the non-breakdown class of clamps, there is no current hogging: all clamps help dissipate the ESD energy. This in turn lowers the effective clamping voltage of the core clamp and also the AC resistance.

Another advantage of several clamps being simultaneously active is based on the savings of the *IR* drops in the interconnect. Normally a core clamp will be placed at certain predetermined intervals along a chip periphery. The idea is that a number of buffers share a core clamp. Thus a zapped buffer needs to pass the current on the core clamp, which is several hundred micrometers (maybe thousands) away. If the clamps are a breakdown device type, they generally "hog" current, and only one of them turns on. Consider the example shown in Figure 3-11, which contrasts the voltage drops due to (a) a single core clamp active and (b) several core clamps active in parallel. The interconnect voltage drop is simply the *IR* drop:

$$V_{\text{int}} = IR \tag{3-1}$$

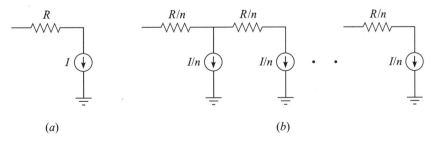

(a)   (b)

Figure 3-11. Benefit of having distributed clamps. Interconnect voltage drop is minimized first by dividing the current in two directions and then by having the current flow in a low-impedance ground plane.

The IR drop can be significant. Consider a 25-μm interconnect having a resistivity of 30 mΩ/□ traveling a distance of 1000 μm. This will have a resistance of 1.2 Ω. Now an HBM 2-kV zap has a peak current of 1.3 A; thus the interconnect loss for an HBM pulse is ∼2.6 V. In the CDM zap case, the current can be as high as 10 A, and in this case the interconnect drop is 12 V. This voltage is in excess of oxide breakdown voltages in most technologies in use today. The heating due to the ESD pulse also increases the resistance; thus the IR drop will only increase [Maloney, 1992, 1993].

However, if the clamps share current, then, instead of placing one clamp in one area, distributing the core clamp into smaller portions such that the total is equal to one large clamp will aid greatly in reducing the interconnect loss. This is shown in Figure 3-11. Here one clamp is broken down into n segments. Assuming equal conduction in all clamps, it is seen in Eq. 3-2 that in such a distributed-clamp scenario the interconnect loss can be reduced by a factor of 4. In reality this improvement may not all be realized, but the branching of current into two paths will immediately lower the drop by half:

$$V_{\text{int}} = \tfrac{1}{2}IR\left(\frac{1}{2} + \frac{1}{2n}\right)$$
$$= \tfrac{1}{4}IR \quad \text{for } n \gg 1 \tag{3-2}$$

In the analysis, a tacit understanding is that V_{ss} is ideal, with no resistance and inductance. An eutectically bonded substrate can approach that. However, when the substrate is not well bonded electrically, the improvement by distributing the clamp may not be a factor of 4. The minimum improvement will be only a factor of 2 if only the closest clamps are considered active. So in practice an improvement between 2 and 4 should be expected.

The model in Figure 3-12 suggests that for packages without a solid backside V_{ss} connection, clamps such as the cantilevered diode and others relying on substrate connectivity need extra substrate taps tied to metal V_{ss} and should be placed immediately next to the clamp in order to maintain their efficacy. Without such substrate tap measures failures have been reported [Johnson, 1993].

3.5. PREDRIVER DESIGNS

In the construction of NBD clamps an inherent assumption has been that the sense and control circuit be considerably faster than an ESD pulse. This is very true in current and future technologies where gate speeds are much less than 200 ps. For the same reason, the predriver is able to influence the behavior of the driver, which has ESD implications. These implications will be considered and design guidelines stated.

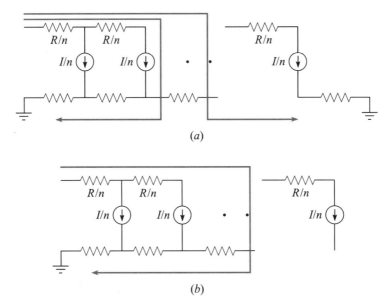

(a)

(b)

Figure 3-12. Benefits of a distributed clamp are diminished if (*a*) good ground plane is not available or (*b*) the other end of the clamp is not referenced to V_{ss}. Between the two cases (*a*) is preferred [Krieger, 1991]; however, good grounding should be encouraged.

Consider Figure 3-13*a*. In this circuit the predriver is powered by the peripheral power supply V_{ccp} and V_{ssp}. The driver consists of two types of NMOS drivers: MNESD, which are solely for ESD protection, and MNIO, which are the active drivers. Now if the pad is zapped with V_{ssp} ground, the transistor Q_1 charges up the V_{ccp} power supply. Since V_{ccc} is independent of V_{ccp}, node N_2 is at a low voltage. Therefore, the predriver inverts the low and drives the NMOS on. This results in the NMOS (MNIO) snapping back and taking up all the current. The NMOS MNESD does not snap back and does not share this ESD stress current with the NMOS MNIO. This effect has been observed even in unsilicided processes where one expects the ballasting action to even out the current flow. Krakauer [1994] reports that in unsilicided technology this effect severely degrades performance (400 V observed as opposed to > 1500 V expected). This phenomenon occurs when there is an odd number of inverting stages tied to the peripheral supply.

An easy solution to this problem is to have an even number of inverting stages tied to the peripheral supply, as shown in Figure 3-13*b*. Here, when the pad is zapped with the V_{ssp} tied to ground, the V_{ccp} charges up by the transistor Q_1. Node N_3 is low as it is tied to the core supplies. However, this time there are an even number of inverting stages powered up. Thus N_1 is also low and the NMOS MNIO does not turn on. This allows the MNESD and MNIO to share

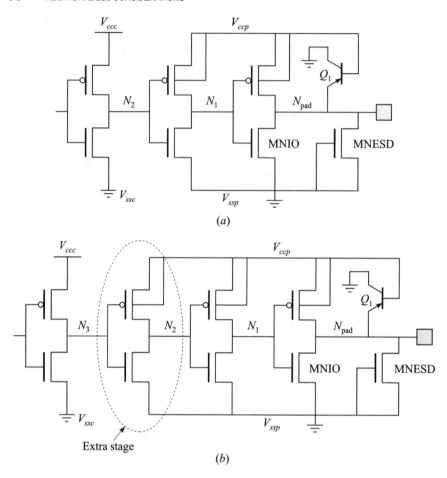

Figure 3-13. Effect of predriver connectivity on ESD reliability of output NMOS: (*a*) odd number of inverting predriver stages; (*b*) even number of inverting stages. It is advisable to have an even number of predriver inverting stages before the driver, as shown in (*b*).

the ESD stress. Significant improvements have been observed [Krakauer, 1994].

The solution above forces both the transistor MNESD and MNIO to see the same gate voltage and thereby have a chance to snap back at similar voltages. It is possible to improve the design further using the concept of the GCNMOS. To recollect it is the phenomenon that when the gate voltage is increased, the snapback voltage of the parasitic NPN decreases. However, if the gate voltage goes far beyond the threshold, the I_{t2} (trigger current for second breakdown) capability also decreases [Amerasekera, 1995; Polgreen, 1992]. So the gate voltage should be modulated to just above the threshold of the NMOS. As seen in the previous example, the high gate voltage caused the lowered ESD robustness when an odd number of predriver stages were used. To take full advantage of even breakdown and current flow and a high I_{t2}, the voltage at the NMOS gate should be high enough to turn on the device but not much higher than that.

A circuit that equalizes the current and preserves the I_{t2} in the NMOS fingers is shown in Figure 3-14. The inverters MP_2 and MN_2 try to pull N_4 high when the pad is zapped and V_{ssp} is low. the NMOS MN_1 is ratioed such that N_4 does not reach the pad voltage but is limited to around V_{tn}. This enables all the fingers of the transistor MNESD to snap back and share the current equally. This is the same principle as a gate-coupled NMOS; however, it is implemented using a ratioed gate (RGNMOS) [Krakauer, 1994].

With the circuit in Figure 3-14, it is possible that if the I/O transients exceed V_{ccc} significantly, then MNESD can turn on and interfere with the I/O operation. However, since the pad already clamps to about V_{cc} plus a diode drop, strong turn-on of MNESD is not expected. Also, the voltage clamp at node N_4 will weaken the turn-on of MNESD.

In previous sections we have seen the effect of external charge built up and then discharged into a pin. Such phenomena have limited access through a pin only. A more prevalent form of charge buildup and possibilities of ESD damage at every node on the chip occur during fabrication. This is in contrast to the earlier case in that nearly every node is affected and the charge access is not limited through the pins. This is considered next.

3.6. ANTENNA DIODE ISSUES

During the fabrication process a number of plasma etching and ion implantation steps are used [Intel, 1991, pp. 9, 10]. these charge up the metal and the polysilicon gates tied to them [Cheng, 1996; Chien, 1997]. If the metal-area-to-polysilicon-gate ratio is low (by a technology-specific ratio), the voltage buildup in the polysilicon gate is tolerable. However, if the collection area is large compared to the polysilicon to which it is tied there is sufficient voltage built up to degrade (and maybe damage) the thin gate oxide. The exact nature of the antenna effect is still being debated [McCarthy, 1990] and new ones are observed [Gabriel, 1996], but most references acknowledge the fact that the charging damage exists. further they generally agree that this is essential to limit this ratio of metal to polysilicon. However, one needs metallization to run signals and power. So limiting the metal wiring is really not a solution. Other alternative solutions are necessary.

Two solutions shown in Figure 3-15 are as follows:

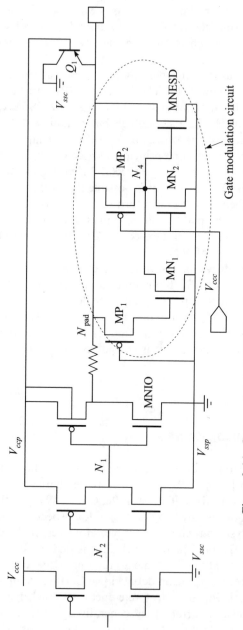

Figure 3-14. Gate-modulated NMOS acting as protection device for I/O.

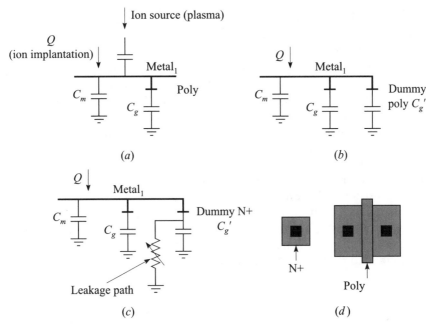

Figure 3-15. Charge sharing during processing can damage gate oxide: addition of (*a*) dummy poly or (*b*) dummy N^+ and (*c*) additional capacitance can reduce the voltage on the gate; (*d*) physical implementation of the ideas show relative areas. Charge source can be capacitively coupled as in a plasma or there can be a direct charge source as during ion implantation.

- Add an extra dummy gate, which can reduce the metal-capacitance-to-gate-capacitance ratio.
- Add extra diffusion (N^+ in P-well) to reduce the voltage buildup in the gate.

From Figure 3-15*d* it can be concluded that the N^+ implementation has advantages over the poly implementation primarily because of the following:

- Smaller area is required (a contact and enclosure of contact) compared to the poly (needs poly and N^+ enclosure of poly on either side)
- Signals may be tied to N^+ as in NMOS drains and sources, in which case no additional device is needed. Here the N^+ should be connected to the gate during the fabrication processes of metal 1 and upward, otherwise an N^+ diode is required.
- The N^+-to-P breakdown usually occurs earlier than the oxide breakdown [Amerasekera, 1994, p. 240].

Thus the N$^+$ helps clamp the overvoltage at the gate. With future gate oxide scaling down the avalanche and oxide breakdown voltages seem to be converging and the benefit of the N$^+$ diffusion will decrease.

The area advantage and ease of routing encourage the use of N$^+$-to-P-epi diodes to reduce gate damage during fabrication. This can occasionally lead to ESD reliability issues. Consider Figure 3-16: two N$^+$ have been placed at minimum spacing and tied to different nodes. Whenever the voltage exceeds the snapback voltage of the parasitic NPN, a large current will flow. If there are external resistors in the path, then they limit the current flow through the snapped-back NPN transistor, usually preventing damage. This is the case when the two neighboring nodes are connected to signal nodes driven by MOS transistors. However, if the node is tied to the external pad or to power supplies, a low resistance exists in the path and large current can flow. Since the areas concerned are minimum-size areas, damage should be expected and in reality it has been observed [Duvvury, 1988].

Solutions to such problems are as follows:

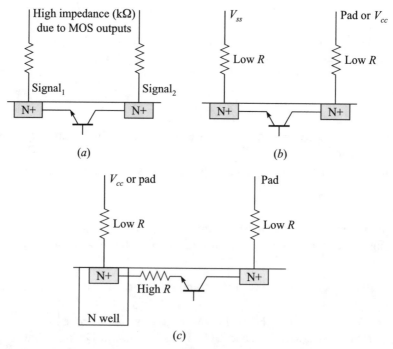

Figure 3-16. Structures such as these occur to protect excess gate damage during fabrication. N$^+$ diffusion is added to reduce the metal-to-gate (plus diffusion) ratio. This structure can be close to a pad or V_{cc} N$^+$. When in close proximity, the parasitic NPN can breakdown and all the charge stored in the pad or V_{cc} node can then flow through, damaging the antenna diode.

- Avoid minimum N^+-to-N^+ spacing (tied to power supplies) during layout, but this is extremely hard to verify. Separating the two N^+ will increase the threshold (V_{sb} is proportional to $L^{0.3}$ [Amerasekera, 1994, p. 237]) but it is again area consuming.
- Add an N-well around such N^+, an area-consuming exercise.
- Use better power supply clamps and large decoupling capacitors so the snapback voltage is never reached. This option can avoid layout difficulties and area penalty and does not need additional verification.

3.7. HOT-ELECTRON INTERACTIONS

The hot-electron effect has been a great concern in modern VLSI transistor design. The phenomenon is the charge injection into the gate oxide near the drain at high electric fields. This excess charge effectively causes threshold shifts near the drain and consequently reduces the current drive of the MOS and therefore slows down the circuit. The root cause of the charge injection is the high electric field near the drain that generates charges from impact ionization. Some of these electrons are injected into the oxide, which leads to the hot-electrons damage. This is a cumulative effect and not catastrophic. Typical solutions utilize lowering the electric field at the drain. This reduces but does not eliminate the effect.

Currently three solutions are employed [Chen, 1988; Sanchez 1989]:

- lightly-doped drain (LDD),
- double-diffused drain (DD), and
- graded junction (GJ).

The LDD is the most commonly used structure (shown in Figure 3-17). Here, a lightly doped drain region is created near the drain that lowers the lateral

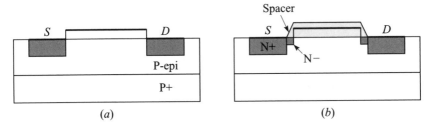

Figure 3-17. Comparison of non-LDD and LDD NMOS. Lightly doped region near the drain causes heat dissipation during an ESD event and increased voltage at the drain, thus causing early ESD failure.

electric field in the drain region, thereby reducing the ionization and subsequent injection. This lightly doped region in turn increases the resistance of the MOSFET, reducing its effective G_m. In addition, it creates hot spots during ESD zapping due to large current concentration and buildup in the electric voltage at the drain (due to larger resistance). Invariably, LDD structures are less ESD robust than the non-LDD options [Amerasekera, 1994, p. 338].

Other circuit-related options exist [Leblebici, 1996; Intel, 1991, pp. 8–10] to lower hot-electron damage in logic circuits:

- nonminimum channel lengths,
- cascaded transistors,
- faster input transitions, and
- increased W compared to the driven capacitive load, thereby reducing the current density.

Two options using nonminimum channel length and cascaded devices to reduce the hot-electron effect are shown in Figure 3-18. Both lower the drain field in each device, leading to a drastically lowered hot-electron effect. For the cascaded option, the hot-electron reduction is typically an order of magnitude better compared to a single device. However, cascading devices roughly double the size of the driver. The extra width required to normalize the driver resistance in turn reduces the current density in the drain, which then helps reduce the hot-electron degradation. The longer channel length in an I/O device also helps increase the snapback voltage of the NMOS, thus helping ESD reliability too.

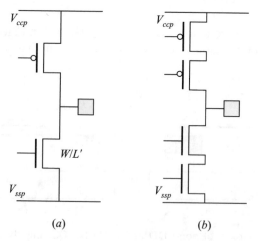

(a)   (b)

Figure 3-18. Two options for increased hot-electron robustness: (a) increased channel length and (b) cascading devices.

The hot-electron degradation is exponentially related to the channel length and the drain voltage, as shown in Eq. 3-3. Further, the drain voltage has the strongest influence in the hot-electron damage to a NMOS device. Thus any drain overvoltage has a major impact on the reliability of the buffer:

$$\delta I = \frac{C}{W} e^{-(k_1 L_{\text{eff}})(k_2 V_{\text{dd}})} \times \text{AF} \times F \times T \qquad (3\text{-}3)$$

where δI is the change in current relative to the undamaged device, C is the capacitance of the toggling node tied to the drain of the NMOS device, W is the width of the device, L_{eff} is the effective gate length, V_{dd} is the drain voltage, and k_1 and k_2 are normalizing constants with respect to L_{eff} and the drain voltage, AF is the activity factor of the expected node toggling (clocking being 1 and other nodes lower), F is the frequency of toggles, and T is the length of time the device toggles (say 15 years).

How does this all tie in with an I/O buffer? As noted earlier, one method to reduce the hot-electron effect is to have fast transition time at the gate of the output device. However, I/O predrivers are usually designed to do exactly the opposite. The predrivers tend to control the slew rates so that signal integrity may be maintained on a network. Thus the I/O buffer is exposed to considerable hot-electron effects for a longer time than compared to a core transistor. In addition an I/O buffer may drive the capacitance of a pad, which is typically in the 3-pF range, and a transmission line. These factors increase the time the output transistors see a high drain-to-source voltage, which causes hot-electron damage.

It has also been observed [Aur, 1988] that the ESD zaps can increase the interface trapped charge in an MOS device, which leads to threshold shifts and reduction in device current. Devices that had been stressed show degraded current compared to completely new devices. Thus an output device that has to meet normal hot-electron requirements needs to have an extra design margin due to the ESD-induced charging. The nonminimum-length output device can fulfill this requirement too.

3.8. LATCHUP ISSUES

A drawback in CMOS technology is its propensity to latch up if proper care is not taken to prevent it. In the I/O area large voltage over- and undershoots frequently forward bias diodes. These lead to large current spikes that raise the substrate potential locally. This potential increase can forward bias parasitic PNP and NPN transistors such that latchup occurs. So in the I/O area it is imperative to place a guard ring in the I/O buffer.

A guard ring structure is shown in Figure 3-19. The N-well can cause electron injection in the P-epi region. An efficient method is needed to pick up these electrons. For this reason an N-well-based guard ring is used. An N^+

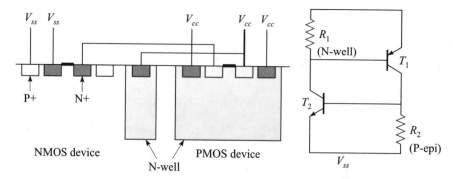

Figure 3-19. Practical implementation of guard rings in the I/O and clamp devices. These devices may occasionally be forward biased and have to be guard ringed.

guard ring will suffice, resulting in reduced guard ring layout area, but it is not sufficiently deep to guarantee good electron pickup. Also the N-well placement forces the PMOS and NMOS to be further separated, degrading the NPN β. Similarly, a P^+ guard ring is placed in the N-well. This picks up the holes in the PMOS N-well as well as degrading the lateral PNP β.

Similarly, the peripheral-to-core-coupling diodes may forward bias during circuit operation. This again can cause a latchup situation A guard ring should also be properly placed. Other diodes that regularly experience conduction are the ones connecting the pad to the supply diodes. Whenever there is a large over- or undershoot, these diodes turn on and provide clamping. In the process they can easily cause latchup if proper guard rings are not implemented. It is the authors' experience that with proper guard rings in place these diodes do not latch up.

To implement a successful guard ring structure, the following guidelines are useful [Weste 1985; Annaratone, 1986, p. 310]:

- Provide a P^+ guard ring around the NMOS and an N^+ guard ring around the PMOS device.
- The substrate taps around the NMOS and the N-well taps for the PMOS should be close to the active sources. this reduces the N-well and P-epi resistance, increasing the charge flow required to forward bias the NPN and PNP transistors.
- Reduce the NPN gain (β) by increasing the distance between the PMOS N-well and the NMOS device.

Lowering the epi thickness will also help reduce latchup. However, this is a process-optimized variable and usually beyond the circuit designer's control.

3.9. SILICON-ON-INSULATOR ESD PROTECTION

The regular bulk CMOS process has handicaps in terms of latchup and extra source and drain capacitance to the body (tied to power). For faster and more robust circuits silicon on insulator (SOI) has been considered. In the roadmap of silicon processes SOI is predicted to be increasingly popular.

The primary difference between the bulk CMOS and SOI is the lower source and drain parasitic capacitance to power supplies, as shown in Figure 3-20. This is achieved by fully depleting the thin active silicon so that the electric field lines are terminated a large distance away. This causes another series capacitance to be effectively tied into the source and drain capacitance, thus reducing the total source and drain capacitance dramatically. The key is to ensure full depletion, which necessitates thin silicon. This reduction in silicon has implications as follows (also see Figure 3-20):

- The reduction in source and drain capacitance considerably enhances circuit speeds.

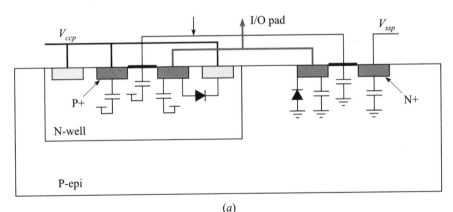

(a)

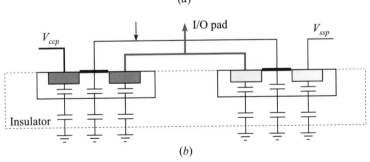

(b)

Figure 3-20. Comparison of (a) bulk CMOS and (b) SOI technology. Parasitic capacitance in source and drain are reduced but the ESD diodes are also absent.

- The PMOS and NMOS are completely isolated so no regular CMOS latchup possibility exists.
- Since the volume of the depletion layer around the PMOS and the NMOS is reduced, there is less chance of external induced currents (radiation), thereby providing higher radiation hardness.
- The ESD diodes in bulk CMOS processes do not exist, thereby complicating the ESD protection mechanism. Both positive and negative ESD zap protection are difficult.
- The thin silicon sandwiched between two insulators has a small thermal mass compared to the bulk CMOS. This reduces the ESD tolerance. To counter this, wider (therefore larger volume) devices need to be constructed.

In the past SOI was employed for exotic systems and ESD reliability was not crucial, as other safeguards could be implemented. However, with SOI technology more widely used, good ESD protection is fast becoming a necessity.

One solution is to use the bulk silicon substrate on which the SOI circuits are built. Conventional ESD protection used in bulk silicon can be appropriately applied to SOI circuits, thereby creating an ESD safe chip. However, this complicates the fabrication process considerably and its benefits should be contrasted with its complexity.

With the lack of good ESD diodes, early research has primarily focused on using the NPN snapback phenomena to provide protection for both the positive and negative direction. There are two properties in the NPN snapback that differ from the bulk CMOS. First, there is no substrate tap; therefore, all the generated impact ionization current as the drain flows through the regenerative junction (source). This lowers the snapback voltage. Second, in a snapback mode, two distinct snapback levels may be observed [Verhaege, 1993]. This NPN action provides protection that can be scaled with the device width ($\sim 12\,V/\mu m$ for an HBM pulse).

A key difference in the SOI technology is that since the silicon volume is small, the device reaches thermal equilibrium rapidly and thermal equilibrium can be observed in even HBM events. Bulk silicon HBM events generate heat that can diffuse into the bulk; therefore the temperatures are lower. Because of the low silicon volume, it has been estimated that the SOI device has only 40% HBM ESD capability as a comparable bulk silicon device [Ramaswamy, 1995].

More recently mapping bulk silicon type ESD protection has been demonstrated for SOI [Voldman, 1996]. These include creating dual-diode input and output protection. In these diodes there is no parasitic PNP action; therefore the forward voltage scales with the number of diodes in series. The resistance also scales as more diodes are placed in series. The small silicon volume forces these diodes to have a large perimeter.

3.10. SUMMARY

In this chapter the following ESD-related factors were discussed:

- Packaging effects on ESD in terms of CDM voltages developed and also the interaction of the ESD pulse with the package. The HBM can interact with the package more than the CDM pulse primarily because of the slower rise times. The larger the dielectric height in a package the smaller the capacitance and therefore the charge picked up by the package.
- The beneficial effect of on-chip and package capacitance in reducing stress is discussed. Further ESD complications of Multi Chip Modules on ESD protection are examined.
- The separation of power planes in the package affects the ESD performance and the choice of ESD protection. If the power planes are isolated on die, then power clamps between the peripheral and core supplies should be provided.
- The main ESD concern in the future should be expected in small dies with small on-chip capacitance and on highly sensitive technologies.
- By distributing the ESD clamps evenly in the I/O area a performance improvement in the IR is observed (in some cases 75% improvement, over a single clamp). This is true only with nonbreakdown clamps which can share the ESD stress amongst them. The breakdown style clamps tend to "hog" all the current to the device which breaks down the earliest.
- The predriver design helps in distributing the ESD stress in NMOS output drivers especially if there are active and dummy ESD legs in the output stage. To avoid nonuniform breakdown and current sharing the pre-driver (or odd numbered) should be connected to the main V_{cc} supply and not the peripheral supply.
- Antenna diodes are added to protect MOS gates during fabrication. They bleed off excess charge during ion processes. However, in poorly ESD protected circuits the NPN snapback of these devices can cause permanent shorts and damage thereby making the chip inoperative. Thus either these diodes should be given a wide room from other N^+ during layout or a better power clamp should be utilized.
- Hot-electron process optimization (LDD) together with silicided drain diffusion creates a hot spot during ESD that limits the device tolerance. One method is to remove the silicidation in the output NMOS drain side, thereby providing a ballast action.
- Due to the slow ramp rates of the predriver (for simultaneous switching output and signal integrity reasons), the output transistor is exposed to high levels of hot-electron damage. To reduce the hot-electron concern, the channel length can be larger than minimum dimensions.

- The latchup hazard in the output devices is best handled by use of guard rings and spacing. These should be combined with proper I/O design where large incoming overdrives are avoided as far as possible.

REFERENCES

[Amerasekera, 1994] A. Amerasekera and C. Duvvury, "The Impact of Technology Scaling on ESD Robustness and Protection Circuit Design," *Proc. EOS/ESD Symp.*, 1994, p. 237.

[Annaratone, 1986] M. Annaratone, *Digital CMOS Circuit Design*, Kluwer Academic, Boston, 1986, p. 77.

[Aur, 1988] S. Aur, A. Chatterjee, and T. Polgreen, "Hot-Electron Reliability and ESD Latent Damage," *IEEE Trans. Elec. Dev.*, 35(12), 1988, p. 2189.

[Chatterjee, 1991] A. Chatterjee, T. Polgreen, and A. Amerasekera, "Design and Simulation of a 4kV ESD Protection Circuit for a 0.5 μm BiCMOS Process," *IEDM*, 1991, p. 913.

[Chen, 1988] K. L. Chen, "The Effects of Interconnect Process and Snapback Voltage on the ESD Failure Threshold of NMOS Transistors," *IEEE Trans. ED*, 35(12), 198, p. 2140.

[Cheng, 1996] H. C. Cheng, T. K. Kang, T. K. Ku, B. T. Dai, and L. P. Chen, "A Novel Technology to Reduce the Antenna Charging Effects During Polysilicon Gate Electron-Cyclotron-Resonance Etching," *IEEE, EDL*, 17(7), 1996, p. 338.

[Chien, 1997] C.-H. Chien, C.-Y. Chang, H.-C. Lin, T.-S. Chang, S.-G. Chiou, L.-P. Chen, and T.-Y. Huang, "Resist Related Damage on Ultra Thin Gate Oxide During Plasma Ashing," *IEEE EDL*, 18(2), 1997, p. 33.

[Cohen, 1997] E. T. Chen, J. Ballard, J. Blomgren, C. S. Branshears, V. Moldenhauer, and J. Patten, "A 533 Mhz BiCMOS Superscalar Microprocessor," *Proc. ISSCC*, 1997, p. 164.

[Dabral, 1993] S. Dabral, R. Aslett, and T. Maloney, "Designing On-Chip Power Supply Coupling Diodes for ESD Protection and Noise Immunity," *Proc. EOS/ESD Symp.*, 1993, p. 239.

[Dabral, 1994] S. Dabral, R. Aslett, and T. Maloney, "Core Clamps for Low Voltage Technologies," *Proc. EOS/ESD Symp.*, 1994, p. 141.

[Duvvury, 1988] C. Duvvury, R. N. Rountree, G. Adams, "Internal Chip ESD Phenomena Beyond the Protection Circuit," *IEEE Trans. ED*, 35(12), 1988, p. 19.

[Gabriel, 1996] C. T. Gabriel and S. R. Naraini, "Correlation of Antenna Charging and Gate Oxide Reliability," *J. Vac. Sci. Tech*, A14(3), 1996, p. 990.

[Intel, 1991] Intel, "Components Quality and Reliability," Intel Corp., 1991.

[Jaffe, 1990] M. Jaffe and P. E. Cottrel, "Electrostatic Discharge Protection in a 4-Mbit DRAM," *Proc. EOS/ESD Symp.*, 1990, p. 218.

[Jain, 1997] A. K. Jain et al., "1.38cm^2 550 MHz Microprocessor with Multimedia Extensions," *Proc. ISSCC*, 1997, p. 174.

[Johnson, 1993] C. C. Johnson, S. Qawami, and T. J. Maloney, "Two Unusual Failure Mechanisms on a Mature CMOS Process," *Proc. EOS/ESD Symp.*, 1993, p. 225.

[Koopman, 1989] N. G. Koopman, T. C. Reily, and P. A. Totta, "Chip-to-Package Interconnections," in *Microelectronic Packaging Handbook*, R. R. Tummala and E. J. Rymaszweski, Eds., Van Nostrand Reinhold, New York, 1989, p. 395.

[Krakauer, 1994] D. Krakauer, K. Mistry, and H. Partovi, "Circuit Interactions During Electrostatic Discharge," *Proc. EOS/ESD*, 1994, p. 113.

[Krieger, 1991] G. Krieger, "Nonuniform ESD Current Distribution due to Improper Metal Routing," *Proc. EOS/ESD Symp.*, 1991, p. 104.

[Leblebici, 1996] Y. Leblebici, "Design Considerations for CMOS Digital Circuits with Improved Hot-Carrier Reliability," *IEEE JSSC*, **3**(7), 1996, p. 1014.

[Maloney, 1988] T. J. Maloney, "Designing MOS Inputs and Outputs to Avoid Oxide Failure in the Charged Device Model," *Proc. EOS/ESD Conf.*, 1988, p. 220.

[Maloney, 1990] T. J. Maloney, "Enhanced P+ Substrate Tap Conductance in The Presence of NPN Snapback," *Proc. EOS/ESD Symp.*, 1990, p. 197.

[Maloney, 1992] T. J. Maloney, "Integrated Circuit Metal in the Charged Device Model: Bootstrap Heating, Melt Damage, and Scaling Laws," *Proc. EOS/ESD Symp.*, 1992, p. 129.

[Maloney, 1993] T. J. Maloney, "Integrated Circuit Metal in the Charged Device Model: Bootstrap Heating, Melt Damage, and Scaling Laws," *J. Electrostatics*, **31**, 1993, p. 313.

[Maloney, 1994] T. J. Maloney, Electrostatic Discharge Protection Circuits Using Biased and Terminated PNP Transistor Chains, U. S. Patent 5,530,612 issued June 25, 1996.

[McCarthy, 1990] A. M. McCarthy, W. Lukaszek, L. Larson, and J. McVitie, "Applications of a New Wafer Surface Charge Monitor," *Proc. EOS/ESD Symp.*, 1990, p. 182.

[Merrill, 1993] R. Merrill and E. Issaq, "ESD Design Methodology," *Proc. EOS/ESD Symp.*, 1993, p. 233.

[Polgreen, 1992]. T. Polgreen and A. Chatterjee, "Improving the ESD Failure Threshold of Silicided *n*-MOS Output Transistors by Ensuring Uniform Current Flow," *IEEE Trans. ED*, **39**(2), 1992, p. 379.

[Ramaswamy, 1995] S. Ramaswamy, P. Raha, E. Rosenbaum, and S. M. Kang, "EOS/ESD Protection Circuit Design for Deep Submicron SOI Technology," *Proc. EOS/ESD*, 1995, p. 212.

[Ramaswamy, 1996] S. Ramaswamy, C. Duvvury, A. Amerasekera, V. Reddy, and S. M. Kang, "EOS/ESD Analysis of High-Density Logic Chips," *Proc. EOS/ESD Symp.*, 1996, p. 285.

[Sanchez, 1989] J. J. Sanchez, K. K. Sueh, and T. A. DeMassa, "Drain-Engineered Hot-Electron Resistance Device Structures: A Review," *IEEE Trans. Electron. Dev.*, **36**(6), 1989, p. 1125.

[Shukla, 1985] R. K. Shukla and N. P. Mencinger, "A Critical Review of VLSI Die-Attachment in High Reliability Applications," *Solid State Tech*, July 1995, p. 67.

[Tandan, 1994] N. Tandan, "ESD Trigger Circuit," *Proc. EOS/ESD Symp.*, 1994, p. 120.

[Verhaege, 1993] K. Verhaege, G. Groesneken, J. P. Colinge, and H. E. Maes, "The ESD Protection Capability of SOI Snapback NMOSFETS: Mechanisms and Failure Modes," *Proc. EOS/ESD Symp.*, 1993, p. 215.

[Voldman, 1992] S. H. Voldman, V. P. Gross, H. J. Hargrove, J. H. Never, J. A. Slinkman, M. O. O'Boyle, T. H. Scott, and J. Deleckl, "Shallow Trench Isolation Double-Diode Electrostatic Discharge Circuit and Interaction with DRAM Output Circuitry," *Proc. EOS/ESD Symp.*, 1992, p. 277.

[Voldman, 1994a] S. Voldman and G. Gerosa, "Mixed-Voltage Interface ESD Protection Circuits for Advanced Microprocessors in Shallow Trench and LOCOS Isolation CMOS Technologies," *IEEE Int. Electron Devices Meeting Proc.*, 1994, p. 277.

[Voldman, 1994b] S. H. Voldman "ESD Protection in a Mixed Voltage and Multi-Rail Disconnected Power Grid Environment in 0.50 and 0.25 μm Channel Length CMOS Technologies," *EOS/ESD Symp.*, 1994, p. 125.

[Voldman, 1996] S. H. Voldman, R. Schulz, J. Howard, V. Gross, S. Wu, A. Yaspir, D. Sadana, H. Hovel, J. Walker, F. Assaderaghi, B. Chen, J.Y-C. Chen, and G. Shahidi, "CMOS-On-SOI ESD Protection Networks," *Proc. EOS/ESD Symp.*, 1996, p. 291.

[Wei, 1993] Y. Wei, Y. Loh, C. Wang, and C. Hu, "Effect of Substrate Contact on ESD Failure of Advanced CMOS Integrated Circuits," *Proc. EOS/ESD Symp.*, 1993, p. 221.

[Weste, 1985] N. Weste and K. Eshraghian, *Principles of CMOS VLSI Design: A Systems Perspective*, Addison-Wesley, Reading, MA, 1985, Chapter 1, p. 60.

CHAPTER 4

INPUT–OUTPUT CIRCUITS

For a chip to communicate with the external world, special I/O drivers and receivers are required. As the internal circuits increase in speed, they process data faster and then communicate it back to the external world for storage, display, or further processing. Also, the number of bits per word has been steadily increasing. Together the longer word size and the fast chips combine to represent a higher bandwidth requirement for the chip interface [Catlett, 1992; Prince, 1994].

Earlier simple I/O drivers and receivers (a starting point is [Annaratone 1986, p. 233]) were sufficient to meet the chip requirements. However, with increasing bandwidth requirements new strategies have to be developed. In this chapter basic as well as novel I/O concepts will be discussed.

The I/O design is integrally related to the transmission line so their properties will be briefly highlighted. Also, concepts of series and parallel termination are useful to I/O implementation and their mapping to standard I/O buffers will be additionally discussed. The I/O scheme timing budget will be reviewed. Also concepts of compensation of buffers for optimum performance and reduced noise will be explained.

4.1. TRANSMISSION LINE PHENOMENA

One practical method of digital communication between chips is the transmission lines on a printed circuit board (PCB). These transmission lines are fast and very economical, which explains their popularity. These transmission lines are thick metal wires (~1 mil) with a strengthened polymer dielectric surrounding it, which in turn is capped by power planes. A simplified PCB construction is shown in Figure 4-1, and it has been contrasted to wiring on the chip. There are two variations of transmission lines possible: microstriplines, which see a split dielectric, and striplines, which see a uniform dielectric environment. For more information the reader is directed to books by Clark [1989] and Flatt [1992]

The transmission line exhibits a velocity of propagation for the signal given by

$$v = \frac{c}{\sqrt{\varepsilon}} = \frac{1}{\sqrt{LC}} \qquad (4\text{-}1)$$

where v is the velocity of propagation in the transmission line, ε is the dielectric constant, and L and C represent the electrical equivalent circuit of the transmission line. This velocity of propagation is solely dependent on the dielectric constant of the insulating material. A common dielectric constant for the PCB material is ~ 4.2, which gives the velocity of propagation close to half the speed of light. This velocity can be calculated to about 174 ps/in., or 69 ps/cm. This holds as long as the transmission line has very low resistive losses. The low resistance is possible in these lines because the copper thickness is ~ 1 mil

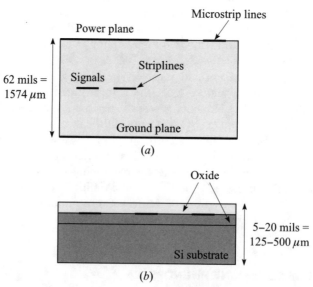

(a)

(b)

Figure 4-1. Construction of (a) a microstripline and a stripline in a PCB and (b) the interconnect on a chip. In a typical PCB multiple signal and power planes are stacked on top of each other, as opposed to the single signal and power plane shown in the schematic. The PCB use an organic dielectric whereas the chips use silicon dioxide as the dielectric.

(25.4 μm) and the widths ∼5 mils. This gives resistance in the low hundreds of milliohms per inch.

A transmission line can be electrically represented as a series of LCR segments, as shown in Figure 4-2. The LC values dictate the velocity of propagation for this line, as shown in Eq. 4-1. Another property of the transmission line is that it behaves like a resistor when a sharp transient pulse is applied to them. This happens because the distributed L and C resist any sharp changes in current or voltage and dominate the input impedance for high frequencies, as shown in Eq. 4-2. Typically in a PCB transmission line the R and G terms are small and can be ignored for a quick Z_0 calculation. This "resistance" is termed the impedance of the transmission line and is really an AC property. The impedance can be given as

$$Z_0 = \sqrt{\frac{\omega L + R}{\omega C + G}} \approx \sqrt{\frac{L}{C}} \tag{4-2}$$

The DC resistance of the line is the summation of all the R elements in series. Although the resistance per segment is small, for long lines this R can be significant. In particular, for I/O schemes relying on large currents the IR drop can be significant.

How sharp should a signal step be to observe transmission line behavior? The transmission line behavior is observed when the rise times of the signal are comparable to twice the time of flight (TOF) (shown in Table 4-1). For a linear first-order system, the rise time is related to the frequency of the signal component, as shown in Eq. 4-3. When the rise times are slower than five times the TOF, then the minimum wavelength (or maximum frequency) is ∼15 times the length of transmission line. This can be seen by substituting the velocity as the length divided by the TOF and the rise time as five times the TOF into Eq. 4-4.

In microwave usage usually a section of a transmission line can be broken down into simplified ideal electrical representation if the current going into the element matches both in magnitude and phase. For this condition, the length of each section should be less than one-tenth of a wavelength. In such a case the dominant $I-V$ characteristics are governed by the ideal electrical characteris-

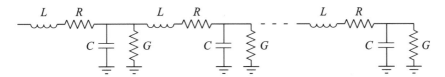

Figure 4-2. Simple transmission line representation using $LCRG$ segments, where L, C, R, and G represent the element values per unit segment of the transmission line: L = magnetic field due to current flow in line; R = resistance the current encounters; C = electric field between line and reference plane; G = resistive losses of dielectric separating line from reference plane.

TABLE 4-1. When to Use Transmission Line Analysis

t_r	Behavior	Comments
$t_r < 2.5t_f$	Transmission line	Significant errors result if transmission lines are not used. Wavelength is less than 7.5 times the transmission line length.
$2.5t_f < t_r < 5t_f$	Mixed	Wavelength is 7.5–15 times the transmission line length.
$t_r > 5t_f$	Lumped	No accuracy is gained by using transmission lines; problem is made more difficult. Wavelength is greater than 15 times the transmission line length.

tics of the R, L, C, and G components of the line. In this case, the complete transmission line is one-fifteenth of the wavelength and it can be approximated as one single lumped section consisting of R, L, C, and G. Further in this section the dominant component is usually the capacitance, and therefore the whole line can be represented as one lumped capacitance element [Bakoglu, 1990, p. 239; Rizzi, 1988].

Using the same argument, when the rise time is smaller than two times the TOF, the transmission line is only one-sixth of a wavelength and a single lumped-element approximation is not as valid any more. For this case a transmission line analysis should be done. The transmission line should be broken down into smaller sections, each under one-tenth of the minimum wavelength (or maximum equivalent frequency) and the sections then connected in series:

$$BW = \frac{0.35}{t_r} \tag{4-3}$$

$$\lambda = \frac{v}{\omega} = v\frac{t_r}{0.35} \tag{4-4}$$

where v is the transmission line velocity and ω is the highest frequency determined by the bandwidth (BW).

When a sharp rise time signal is launched in a transmission line, it propagates to the end of the line as a current and a voltage wave. At the end of the line, depending on the termination, a reflection is developed. This reflection has to be consistent with Kirchhoff's current and voltage laws.

If the line is open, the current has to be completely reflected back, as it is not longer possible for the current to carry on ahead. So a negative current (equal current going in the reverse direction) needs to be developed. For this to happen, the voltage needs to double such that the reflected current is equal to the incident current but only reversed in direction. If the termination is a short circuit, then the voltage is zero and a negative voltage is reflected back. If the transmission line is perfectly matched to the termination, then the current and voltage are satisfied at the end of the line and no reflection is developed. The reflection is usually described in terms of a reflection coefficient Γ and is given by

$$\Gamma = \frac{Z_l - Z_0}{Z_l + Z_0} \tag{4-5}$$

where Z_l is the termination impedance. The effective reflection coefficients are presented in Table 4-2.

The voltage step on a transmission line can be calculated as

$$V = IZ_0 \tag{4-6}$$

where I is the current in the transmission line. This equation is helpful in determining the current required to induce the voltage step and in turn to calculate the driver size. An alternative method to derive this current is to look at the charge required to bring the transmission line capacitance to the required step height. This can be expressed as a function of the transmission line capacitance (c), the distance traveled in a second (v), and the voltage step required, which is shown in the equation

$$I = \frac{dQ}{dt} = \left(\frac{\text{capacitance}}{\text{cm}}\right) \times \frac{\text{cm}}{\text{s}} \times (\text{voltage\_step})$$
$$= C \times v \times V_{\text{step}} \tag{4-7}$$

For a simple CMOS system (rail-to-rail swing) V_{step} is half of V_{cc}.

It is interesting to contrast the above-discussed off-chip transmission lines with the on-chip signals that use thin metal lines to communicate. The TOF aspect has already been discussed in the previous paragraphs. Now the effects of metal resistivity are considered. The IC interconnect lines have resistivity of $\sim 50\,\text{m}\Omega/\square$. Thus a line $1000\,\mu\text{m}$ long and $2\,\mu\text{m}$ wide has a resistance of $25\,\Omega$. Here the capacitance is also larger and the inductance small compared to a PCB line. The resulting impedance is therefore small. Also consider the propagation delay based on a transmission line model, which is approximately

TABLE 4-2. Effect of Termination on Reflections on a Transmission Line

Termination	Γ	Comments
Z_0	0	Behaves as an infinite line; no reflection
Short	-1	Voltage collapses to zero
Open	$+1$	Voltage doubled at open

3.3 ps (1000 μm at a speed of $c/\sqrt{\varepsilon}$) which is negligible. If the DC resistance of the complete line exceeds the characteristic impedance, then the transmission line behaves more like an *RC* line than like a transmission line [Bakoglu, 1990, p. 244]. These IC interconnection lines have *RC*-dominated delays and so behave more like distributed *RC* lines than like transmission lines. The characteristics of a lossy and lossless line are compared in Figure 4-3. It is clearly seen from Figure 4-3 that for full transmission-line-like signaling *Rl* should be much lower than Z_0. A PCB line that has low resistivity (~ 100 mΩ/in.) can traverse tens of inches ($Rl \sim 1\,\Omega \ll Z_0$) and still have a sharp signal appear at the end of the line.

Consider on-chip communication. When the distance is small (thousands of micrometers), the signal delay is comparable to gate delays and the total delay is in the tolerable range [Long, 1990]. However, with increasing die size and faster cycle times (implying faster rise and fall times and smaller gate delays), the same interconnection technology begins to resemble a highly lossy transmission line. A low-loss transmission line would indeed speed up communications within a chip; however, current technologies do not support such transmission lines. In the future, with shrinking timing budgets and larger die, the on-chip transmission line effects will become increasingly important.

The primary properties of transmission lines are the characteristic impedance and the velocity of propagation discussed in this section. In the next section the relationship between I/O drivers, terminators, and transmission lines will be examined.

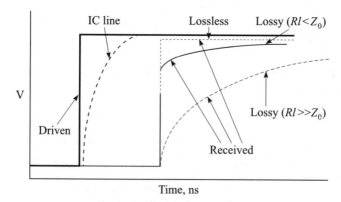

Figure 4-3. Effect of resistance on transmission line performance. For very lossy lines ($Rl \gg Z_0$), the *RC* effect dominates and the propagation delay is no longer set by the dielectric constant alone. For the IC line, the *RC* effect dominates the line response.

4.2. SERIES AND PARALLEL TERMINATION

The driver, the transmission line, and termination matching are key to clean signaling. Transmission lines that are not well terminated suffer multiple reflections leading to an unsettled net for considerable time. These reflections interfere with signaling as new data will be affected by the remnants of the previous data that have not settled down. This is referred to as intersymbol interference (ISI). Thus, the TOF (length) of the transmission line and the termination define the data rate sustainable in a transmission line. Two types of termination are common—series and parallel—and are illustrated in Figure 4-4.

In a series-terminated network, the wave is generated at the driver and propagates down the transmission line to the end, then is reflected back, and finally is terminated at the driver end. This can be seen with the help of Figure 4-4c, where the series-terminated line is driven by a driver of impedance Z_0. At the initial driving edge, the driver voltage is halved at the voltage division between the transmission line impedance and the driver. The current is correspondingly $V_{cc}/2Z_0$. This voltage and current wave propagates to the end of the line. The end of the line is an open circuit; therefore, the current cannot proceed any further. There is no source or sink; consequently all the current must be turned back (Kirchhoff's current law states that the current at a node must be balanced). This is shown as the negative current at the far end. The far-end node is at $\frac{1}{2} V_{cc}$, corresponding to the impressed voltage at the driver wave. To reverse the current completely in a transmission line with impedance Z_0, it needs an additional $\frac{1}{2} V_{cc}$ step over the existing voltage. Therefore the voltage doubles at the far end. This voltage wave comes back to the driver end. At arrival, the net current at the driver falls to zero and the voltage doubles to the V_{cc} level. Now both ends of the driver are at V_{cc}, so no further current flow occurs in the resistor, and the bus settles to a V_{cc} level.

The case where the far end of the line is short circuited can also be analyzed using Kirchhoff's voltage law at the far end. Initially the near-end current is $V_{cc}/2Z_0$ and a half V_{cc} voltage step is impressed. However, at the far end the voltage is zero; therefore, a negative-voltage wave is developed that is reflected back to the driver. This also impresses a current of $V_{cc}/2Z_0$, which adds to the previous current wave, making the current V_{cc}/Z_0. When the voltage- and current-reflected waves reach the near end, the driver would supply V_{cc}/Z_0 current (because the voltage across is V_{cc} and its impedance is Z_0), which is the same as that which exists in the transmission line. Therefore Kirchhoff's voltage and current laws are satisfied and no reflections are developed. If the source termination matches the transmission line, then there is no further ringing on the line. Otherwise, part of the wave is repeatedly reflected back each time the energy of the wave attenuates until the net settles. This scheme is generally popular with the slower CMOS-type I/O or where the net is small. Once the net settles, there is no further power dissipation. For faster I/O with long nets either the source impedance should be close to the transmission line impedance or a parallel termination strategy opted.

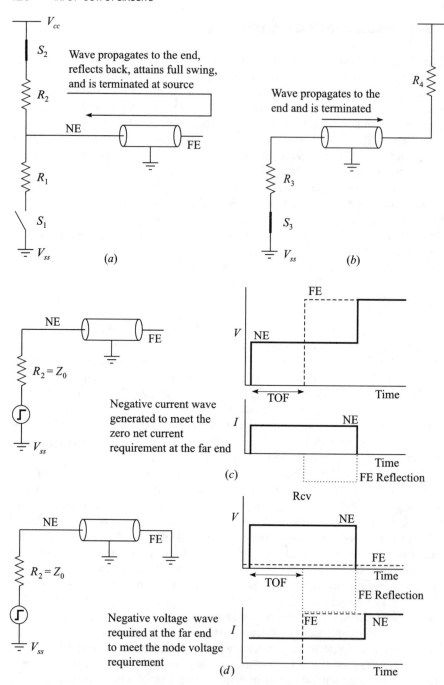

Figure 4-4. (*a*) series and (*b*) parallel termination scheme: line behavior with (*c*) open circuit at the far end and (*d*) short circuit at the far end.

In a parallel termination, the driver launches a wave that transmits to the end, where it is terminated. Here, with proper selection of the impedance, the net may settle faster as termination is provided at both ends of the line. This type of scheme is preferred in faster I/O types, like ECL and bipolar. More recently the MOS has started rivalling BJT I/O speeds, and similar schemes can be applied there also. One such example is the Gunning transceiver logic (GTL) scheme [Gunning, 1992]. These terminations have power dissipation in one state (low or high) even when they settle. In the other state (high or low) they do not have DC power dissipation.

It should be noted that many different I/O families exist, like ECL, CMOS, GTL, and open drain (bipolar and CMOS). Here we shall examine the two popular ones, namely CMOS and GTL.

4.3. CMOS I/O

Complementary MOS I/O has been the workhorse of the computer technology. Put simply, the driver is one big inverter. However, this inverter must meet certain other criteria.

- crowbar current control,
- slew rate control for the predriver,
- high impedance or tri-statable,
- driver size and output resistance control, and
- propagation delay from clock to output control.

Crowbar current is the short circuit current in an output driver caused by both the NMOS and PMOS being partially on when the output node toggles. This crowbar is similar to the one encountered in core logic circuits, only the magnitude is larger and some additional options exist to reduce the crowbarring. However, in a practical high-speed driver, some crowbarring has to be traded off in favor of speed (faster clock to output) and slew rate control.

The slew rate (i.e., rate of change of current or voltage) of the driver is controlled usually by adjusting the predriver circuit. The predriver is that stage of logic between the core circuits and the final output driver that adjusts the timing and the drive to the final I/O output stage such that it meets the required I/O specification. Also the predriver is a large buffer that interfaces the small core logic latch to the large capacitances of the driver. In Figure 4-5, the NAND and NOR gates constitute the predriver. Fast predrivers reduce the propagation time for the data to flow from the chip core to the output driver, but they generate sharp current spikes. When a number of buffers switch simultaneously, this spike can seriously inject noise into the power supply. Thus, it is essential to trade off between noise sensitivity and slew rates and propagation delays. This may require slowing down the predriver.

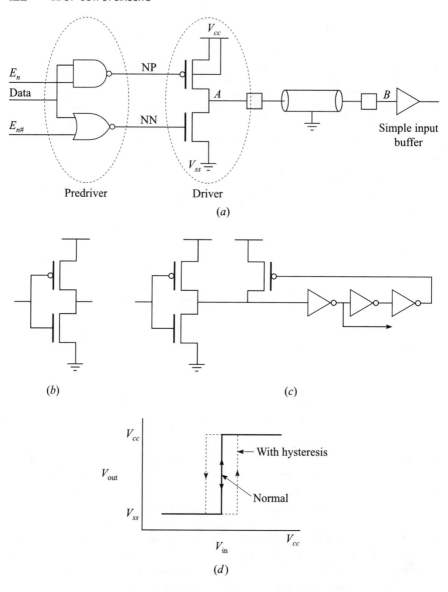

Figure 4-5. Simple CMOS buffer, showing (*a*) a simple predriver, driver, and input receiver combination; (*b*) an inverter as an input receiver; (*c*) an input receiver with hysteresis; and (*d*) the input buffer characteristics.

In a typical design, usually slew rates are adjusted for the slowest condition of the design. The predriver can be slowed such that it can charge/discharge to a specified degree within a data window period. This is illustrated in Figure 4-6. If the predriver is too slow, then there will be a data-dependent jitter. This

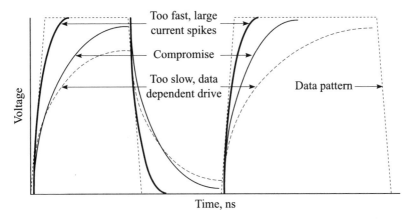

Figure 4-6. Predriver sizing has to comprehend both the noise generation due to rapid switching of the buffers and the data-dependent jitter. The data-dependent jitter occurs when the predriver output voltage does not reach the rail within one data window, allowing different pulse width data streams to charge the driver gate to different voltage levels. The different voltage leads to variation in driver impedance and also variation in the starting point for the next cycle, both contributing to the jitter.

occurs because of the nonlinear predriver output, which behaves like an RC circuit. Thus for complete data independence, the RC time constant must be much smaller ($\ll 3$) than the data window to reach close to V_{cc} or V_{ss} rail voltages. If this is attempted, it may result in the fast-driver case having an extremely sharp output. The fast condition will draw larger currents, have sharper transients, have more simultaneous switching noise, and generate larger cross talk, as shown in Figure 4-6. Based on the fast results, the design for the slow condition may have to be revised. The control of data dependency (in the slow condition) and the control of noise and current (in the fast condition) are the two constraints that have to be balanced. These iterations allow a reasonable simultaneous switching output (SSO) noise by reducing the dI/dt of the outgoing signal. However, the time to clock out the data (T_{co}) is increased and this can cause timing issues. Therefore a trade-off has to be made between SSO, the data-dependent jitter, and T_{co}. Practical circuits to implement the slew rate control will be discussed later in the chapter.

It was shown earlier how good source impedance matching reduces the reflections in a transmission line. By having slower turn-on and turn-off of the drivers, in effect the output impedance of the driver is altered. If peak voltage levels for high- and low-frequency patterns are sufficiently different, the output impedance is different, leading to different reflection characteristics. These will cause an additional data-pattern-dependent ISI.

The driver size also affects the final slew rate of the driver. For a simple source-terminated system, the idea is to have enough current to launch a $\frac{1}{2}V_{cc}$ wave into the transmission line. When it reaches the receiver, this wave will double to reach a full V_{cc} level. This should be done for the slowest condition, where the driver is the weakest. As earlier, in the fast condition the driver becomes strong, and ~50% larger currents are not uncommon. This excess current will cause severe over- and undershoots at the receiver. At the driver end the consequence of the stronger drivers are worse SSO noise and larger power dissipation.

The input end of a CMOS I/O scheme has a simple CMOS inverter as a receiver, as shown in Figure 4-5c. In a CMOS I/O, the voltage swings from rail to rail. This allows a healthy noise margin for the input buffer. The noise margin can be defined similar to core circuits. In addition, sometimes a hysteresis is placed in the input buffer. This can be done by resizing the transistors. For example, consider a simple balanced inverter that may have a P and N size of 20 and 10 μm. Now to create a hysteresis of, say, 10%, the N size is retained as 10 but the PMOS is split into two sizes, 18 and 4 μm, the smaller transistor being used to create the hysteresis. This is shown in Figure 4-5c.

The low noise margin is the difference between what the driver can drive low and the voltage that the inverter considers low. Similarly the high noise margin is the difference between the lowest value the output driver will drive when high and the lowest value recognized by the input inverter as high [Weste, 1985, p. 51]. These are illustrated in Figure 4-7 and by the equation

$$N_{ml} = |V_{ilmax} - V_{olmax}| \qquad N_{mh} = |V_{ohmin} - V_{ihmin}| \qquad (4\text{-}8)$$

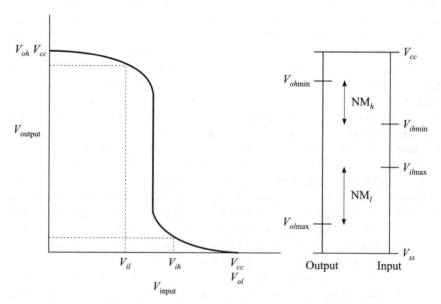

Figure 4-7. Noise margin calculation for a CMOS I/O scheme [Weste, 1985, p. 51].

There are some major differences between the core and I/O noise margins:

- The I/O signal sees the effect of ground bounces generated by the receiver or driver, which are much smaller (or absent) in the core.
- The I/O signal is usually not as well behaved as a core signal. The I/O signal usually has severe undershoots or overshoots and bounce backs, each affecting the noise margin (more on this in Chapter 6).

It was tacitly assumed that a driver could place a step of $\frac{1}{2}V_{cc}$ to have a resistance equal to the transmission line. This is easily done if an ideal resistor is available at the source. However, in reality an NMOS device has to generate the step and act as the termination, and a compromise must be made. This is explained with the help of the load line shown in Figure 4-8.

Assuming the transmission line is at V_{cc} from the previous state and needs to be pulled low, the initial state is that there is no I_{ds} current and the voltage is V_{cc}. This is plotted on the X axis. Similarly, if V_{ds} is set to zero, then $I_{ds} = V_{cc}/Z_0$. This is plotted on the Y axis. The result is a load line for the

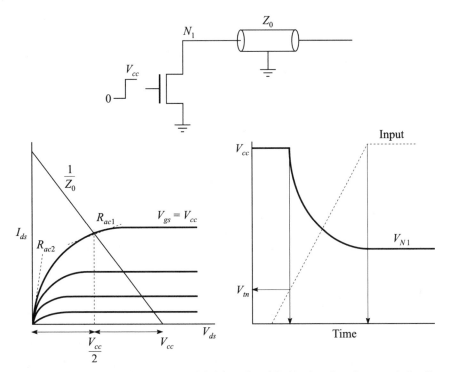

Figure 4-8. Interaction between NMOS driver (nonideal resistor) and transmission line showing (*a*) a load line between the NMOS and the transmission line that (*b*) shows a temporal view between the linearly ramped input voltage and the nonlinear output.

transmission line with a slope $1/Z_0$. The transistor I_{ds} as a function of V_{ds} and V_{gs} are superimposed. The point to generate a step of half V_{cc} is shown as the intersection point between the transmission line (Z_0) load line and the transistor characteristics. If the transistor characteristics do not match the desired half V_{cc} point, the transistor can be scaled up or down such that at the intersection half V_{cc} is realized. It should be noted that the transistor small-signal resistance (R_{ac1}) may be different from the ideal Z_0. This is the point of operation when the wave is driven.

When the reflection comes back, V_{ds} is close to zero. When in the linear region (R_{ac2}) (close to the origin V_{ds}) the output resistance can be written as

$$R_{out} = \frac{L}{W \mu_n C_{ox}(V_{gs} - V_{tn})} \tag{4-9}$$

and

$$R_{ac1} > R_{ac2} \tag{4-10}$$

The wave sees a termination of R_{out} that may be off from the ideal Z_0. This will cause reflections and will in turn lead to an unsettled bus that in turn will lead to ISI. Therefore, several iterations may have to be made to find a good compromise between the initial launch current and the termination resistance. The variation in R_{ac1} and R_{ac2} is smaller if the driver is able to drive the initial wave closer to the linear region than to the saturation region of the NMOS device.

4.4. GUNNING TRANSCEIVER LOGIC (GTL) I/O

The previous section discussed a source-terminated CMOS scheme. Next a parallel-terminated GTL scheme is considered [Gunning, 1992]. In the parallel termination scheme, the line is terminated at the far end. The GTL I/O is a special implementation of this scheme in that it is a low-voltage swing signaling method, as shown in Figure 4-9. The GTL scheme is similar to an open-drain scheme. An active pull-down driver is required whereas the pull-up action is performed using a termination/pull-up resistor. Since there is termination at the end of a net for long buses, this is a suitable scheme in that it settles the bus quicker than a source-terminated net. The low-voltage swing also imposes constraints like lower noise margin and requires special differential receivers. The driver consists of a pull-down device in the form of a large NMOS device.

This scheme is a low-voltage swing scheme with a high at V_{tt} and a low of $\sim 0.4\,\text{V}$. Typically for low-voltage swings, a differential amplifier is used as an input receiver to maximize the noise margin. Thus one input of the differential amplifier is fed to the I/O voltage and the other input to a reference voltage that is typically a fraction of V_{tt}. This differential amplifier is a simple device

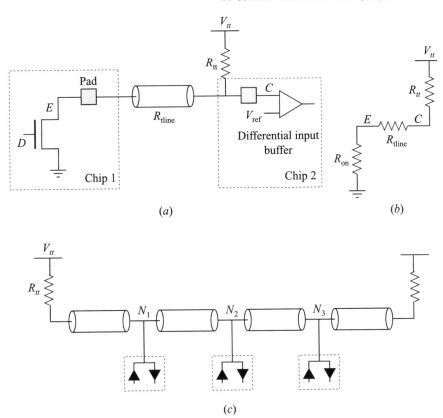

Figure 4-9. Simple GTL buffer showing (*a*) the pull-up resistor/termination, a differential input receiver; (*b*) the DC equivalent model for V_{ol} calculations; and (*c*) a practical GTL bus.

similar to the one shown in Figure 4-10. Here the PMOS transistors are the loads and a current-biased NMOS is used to bias the whole differential amplifier. Consider node B to be higher than the node A voltage. This steers the current into $NMOS_1$ consequently causing node C to go low. This low is not V_{ss}, and a CMOS stage must be inserted to restore this signal to CMOS levels [Bakoglu, 1990, p. 162].

The driver size can be calculated based on DC considerations. Knowing the termination resistor, the DC resistance, and the length of the transmission line, the DC resistance of the driver can be calculated as

$$V_{ol} = V_{tt} \frac{R_{on}}{R_{on} + R_{tline} + R_{tt}/2} \tag{4-11}$$

$$V_{oh} = V_{tt} \tag{4-12}$$

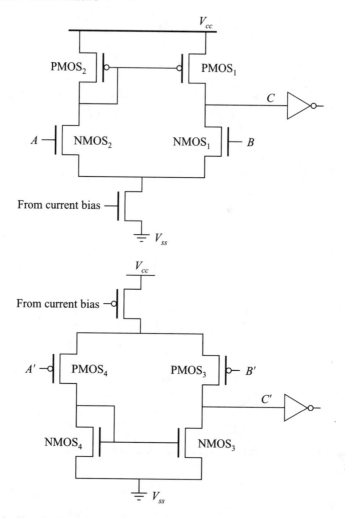

Figure 4-10. Simple differential input buffer. The NMOS are the sensing transistors and the PMOS are the loads. The CMOS inverter is required to restore the final output of the differential amplifier to a CMOS voltage swing.

Here V_{ol} at the driver is set by V_{il} at the receiver (see Figure 4-11). When pulling low, there is a DC current flow and the voltage at the driver differs from the one at the receiver by the IR drop of the transmission line. Thus the V_{ol} of the driver must comprehend this drop. Keeping this in mind, V_{ol} can be recalculated as

$$V_{ol} = V_{tt} \frac{R_{on} + R_{tline}}{R_{on} + R_{tline} + R_{on}/2} \tag{4-13}$$

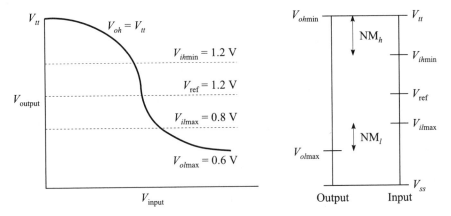

Figure 4-11. The GTL has a reduced swing going from V_{tt} to a low but not to ground. Usually a differential amplifier is used to detect the signal level at the receiver. One end of the differential amplifier is tied to a reference voltage V_{ref} [Intel, 1996]. These voltages are different from the initial GTL description [Gunning, 1992] where $V_{tt} = 1.2\,\text{V}$, $V_{ref} = 0.8\,\text{V}$, and $V_{olmax} = 0.4\,\text{V}$.

$$V_{oh} = V_{tt} \tag{4-14}$$
$$V_{ref} = \tfrac{2}{3}V_{tt} \tag{4-15}$$

It is interesting to contrast the GTL signaling scheme to the CMOS scheme. There is a constant power dissipation when the pull-down buffer is on, whereas in the CMOS scheme there is no power dissipation once the bus has settled. Another difference is the distribution of termination/pull-up V_{tt} power at the termination and the V_{ref} signal to each on-chip receiver, whereas no external termination is required in the CMOS scheme and generally no V_{ref} signal needs to be distributed. These differences are summarized in Table 4-3.

Another subtle effect with the GTL scheme is that it has a different effective termination for the low pull-down and the high pull-up states. When pulling low, the net is terminated at both ends and it has a pull-down at the driver location. However, when the net has to be driven high, the driver releases the

TABLE 4-3. Comparison of GTL and CMOS I/O Schemes

Parameter	GTL	CMOS
Voltage dependent on technology	No	Yes
V_{cc} pads required for I/O	No	Yes
Pull-up external termination	Yes	No
Noise margin	Lower	Higher
Power dissipation (steady state)	Yes (when low)	No
Current requirement	High	Medium

bus, and the only termination is at the end of the net provided by the pull-up termination. This differential effective termination leads to slightly different rising and falling edge settling times and duty cycles. For schemes requiring very similar rising and falling edges, this can be a considerable issue.

In both the GTL and CMOS I/O schemes one can improve the bus performance and increase the noise margins if a differential signaling scheme is adopted in favor of a single-ended scheme. In the single-ended scheme, the signal is compared to a stable reference voltage or the device sizing (as in a CMOS inverter receiver) determines the trip point. In a differential scheme, a differential receiver compares two incoming signals, the data and the inverted data, and uses these to determine the receiver state. These differential signals reduce the common mode noise and provide a much sharper definition of the data edge. Both of these lead to enhanced bus performance and robustness. However, the price is doubling in the number of wires and on-chip buffers [Rein, 1996].

In the last two sections, practical examples of source and parallel terminated buses were provided. Design considerations of their DC properties, such as noise margins, power requirements, and basic topology, were discussed. In the next section the timing considerations are examined.

4.5. BUS TIMING CONSIDERATIONS

For a bus to work correctly, the DC voltages and currents and AC timing need to be satisfied. The main criterion in meeting the timing requirement is to ensure that the right data are available at the right clock edge of the receiver to enable correct capture of the data. There are two major methods to provide this clock for data transfer and capture:

- synchronous signaling, which uses one system clock to both send and receive data, referred to here as a common clock, and
- asynchronous signaling, in which the data and a clock are sent by the driver to enable the data capture at the receiver using the driven clock. Three different clock domains exist: the driver chip, the I/O clocks, and the receiver clock. This scheme is referred to as a cotransmitted clock (CTC).

These schemes are discussed in the following sections.

4.5.1. Common Clock Transfers

In the common clock scheme a common reference clock source is used to clock each chip in the system. Consider the three-chip system shown in Figure 4-12 utilizing a common clock signaling scheme. Each chip may have a phase-locked

Bidirectional data bus

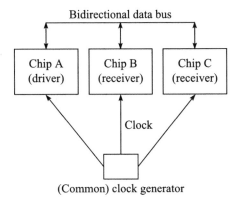

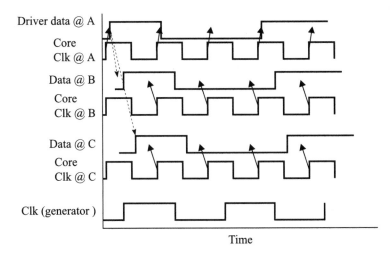

Figure 4-12. Schematic of common clock I/O protocol. Here a central system clock is used to drive the data at the driver and to latch the data at the receiver. There is only one fundamental I/O clock.

loop (PLL) for clock recovery or use the clock directly as received. The data are driven from chip A at the chip A clock edge. The data are latched in at the clock edges of chips B and C at their respective chip clocks. All the chip clocks are referenced back to the common clock source.

The common clock scheme works well when there is sufficient time for the data to be stable before (setup) and after (hold) at the receiver clock edge. Consider the setup time. To meet the setup time, the data have to come before the clock at a receiver, implying the propagation time (including the driver delay and transmission line delay) must be smaller than a clock cycle. Similarly, to meet the hold time, the data must be stable for a minimum period after the clock edge. The hold time dictates that some minimum delay (at the data driver and the transmission line delay) needs to be met.

A more detailed structure of the signaling scheme is shown in Figure 4-13, and the delays through the individual components in the loop are also shown. In Figure 4-12, a synchronized clock generation is shown between all clock edges. In reality, depending on the chip conditions, a chip clock may be skewed either positively or negatively with reference to the common clock generator. With this additional consideration the setup and hold times need to be examined.

Again consider the setup time requirement at the receiver. The worst-case setup timing is obtained when the clock arrives late on driver chip A, the driving flip flop is slowest, and the driver, the transmission line path (to chip C), and the receiver buffers all have maximum delays. In addition to these maximum delays, the clock on chip C comes early. This leaves the minimum amount of time for signal propagation and sets the maximum distance between chips A and B. This setup requirement is given as

$$t_{su} \leq t_{cyc} - (t_{poa,max} + t_{ja,max} + t_{ff,max} + t_{d,max} + t_{t,max} + t_{r,max} + t_{jb,min} + t_{pob,min})$$

$$(4\text{-}16)$$

where each delay is associated with the element marked in Figure 4-13, t_{su} is the minimum setup time required for the receiver latch, and t_{poa} and t_{pob} are the DC phase offsets in the PLL on chip A and the PLL on chip B; t_{ja} and t_{jb} are the PLL jitters.

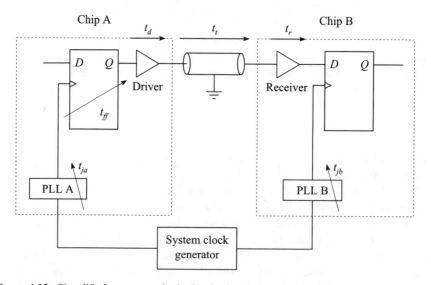

Figure 4-13. Simplified common clock circuit showing typical path from driver latch to receiver latch.

Similarly for hold time consideration, the clock arrives early to the latch on chip A. The driver, the transmission line, and the receiver are all fast, and the clock on chip B comes late. This creates the worst-case hold time such that the data change before being captured on chip B, the nearest chip. Thus the nearest chip should be placed at a minimum distance such that the data change only after a hold time, after the clock edge on chip B. This is shown in the equation.

$$t_{\text{hld}} > (t_{\text{ff,min}} + t_{d,\text{min}} + t_{t,\text{min}} + t_{r,\text{min}} + t_{\text{pob,max}} + t_{\text{jb,max}}) - (t_{\text{poa,min}} + t_{\text{ja,min}})$$

$$(4\text{-}17)$$

where t_{hld} is the minimum hold time required for the latch at the receiver.

Very often, the setup time or the hold time is not met and iteration is necessary. A simple design methodology for the common clock system is shown in Figure 4-14. First, the system topology is chosen based on parameters like speed, number of chips, cost, and packaging. Then the closest chip distance is evaluated using the hold time consideration and the setup time is evaluated using the longest path consideration.

Some of the criteria used to evaluate circuit functionality are shown in Table 4-4. Generally the slowest process, low voltage, high temperature causing largest delays, the delayed clock at the driver, and an early clock at the receiver for the furthest chips determine the setup criteria. The hold time criteria are defined by the fastest conditions, shortest paths, fastest voltages, and fastest processes.

Simple solutions to correct hold time errors are to increase the data path length between the problem chips or reduce the clock delay to the receiver latch (so that it comes early), as illustrated in Figure 4-15. The data path length causes the data to be delayed compared to the clock, thereby allowing the clock

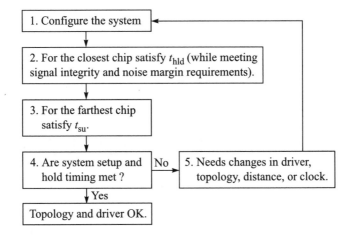

Figure 4-14. Flow chart showing common clock design cycle.

TABLE 4-4. Worst Case Scenario for Common Clocking Scheme

Criteria	t_{setup} Consideration	t_{hold} Consideration
Driver–receiver position	Farthest	Nearest
Clock (driver/receiver)	Late/early	Early/late
Driver	Slow	Fast
Receiver	Slow	Fast
Voltage	Low	High
Temperature	High	Low
Path	Longest	Shortest
Transition time	Slowest	Fastest

Note: The furthest chip receives the data at its slowest timing, and the nearest chip receives data at its quickest timing.

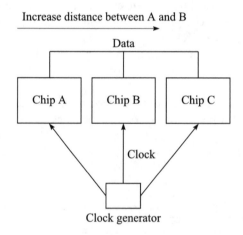

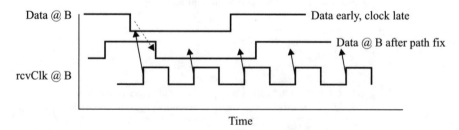

Figure 4-15. Fixing hold time violation at chip A by increasing delay between chips A and B. In this scenario delay to chip C will also increase and setup time will need evaluation.

to catch the data, or the data can be delayed using delay elements such as inverters and delay lines. Both will affect the setup time. The clock to the receiver latch can be speeded up by shortening the system connection or the internal on-chip distribution. This correction may allow no setup penalty but may have consequences when the receiver chip becomes the driver.

4.5.2. Cotransmitted Clock (CTC)

The common clock scheme has some inherent limitations in terms of speed, primarily due to TOF considerations compounded with the PLL phase errors of the driving and receiving chips. These chips become highly visible if the bus is long, leading to large propagation time, or if there is a large chip-to-chip clock skew, both of which can be partially alleviated by a simple method called the CTC scheme [Rambus; Dally, 1996, p. 6]. In this scheme, the data and clocks are generated and sent by the driver as shown in Figure 4-16. The

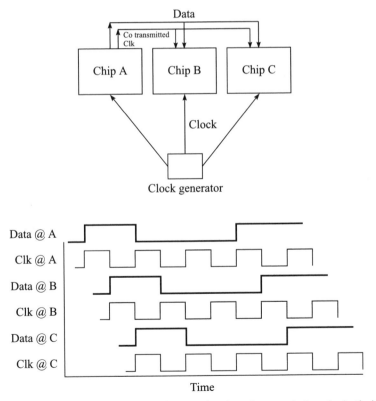

Figure 4-16. Simplified cotransmitted scheme, showing data and the clock timings. Note that a precise known timing interval exists between the data and its latching clock at all receivers.

receiver delays the clock 90° to latch in the data [Dally, 1996, p. 6; Guo, 1994; Reese, 1994]. Theoretically this avoids the penalty of TOF, PLL phase errors, and other common mode errors between the data and the strobe. The increase in I/O performance is significant.

However, there are now three clock domains, the driver core, the receiver core, and the I/O clock, as indicated in Figure 4-17. The I/O clock is derived from the driver clock so it is not independent. However, the data received with respect to the I/O clock have to be aligned back to the receiver core clock domain. When the incoming data are asynchronous to the receiver core clock, the data alignment can be done using shift registers [Stone, 1982; Dally, 1996, p. 41]. This entails additional circuits and delays. Overall, this scheme has a higher latency for the first data transferred. However, in a burst transfer, a pipelined data transfer occurs, and this latency penalty can be offset by the speed of the transfer of the successive bits.

In another variant of this CTC scheme, two sophisticated PLLs in each chip lock into two clocks sent out by one common reference clock, only in opposite directions, as shown in Figure 4-18. This action allows the clocks to lock in and capture the clock skew down a line dependent on the distances they are spaced apart. Then the driver uses one clock to drive the data, and the receiver uses its received clock to latch in the data [Lee, 1994]. The receiver clock is selected such that the data and clock directions are the same. This is shown in Figure 4-18. Theoretically this allows cancellation of the TOF for the data from the setup and hold equations used for a common clock scheme because the clock is also traveling an equal distance as the data. This scheme is very sensitive to clock jitter and differential skews (between data and clock), and very sophisticated PLL designs are required. If executed correctly, this scheme can significantly speed up data transfers.

Theoretically, the data, the receiver, and the driver clock relationship can be maintained and the differential timing scheme works just fine; however, certain elements make the data and clock timings unpredictable and have to be guarded against. The primary components to this unpredictability are the following:

- driver T_{co} (clock-to-output) variations;
- capacitance at pad variations;
- L_{eff} variations;
- interconnect path difference on pads, chip, and PCB;
- clock recovery and generation mismatches;
- data-dependent jitters;
- duty cycle variations;
- noise on chip: ground bounce due to SSO and core circuit operation; and
- receiver reference voltage variations and receiving input circuit offsets.

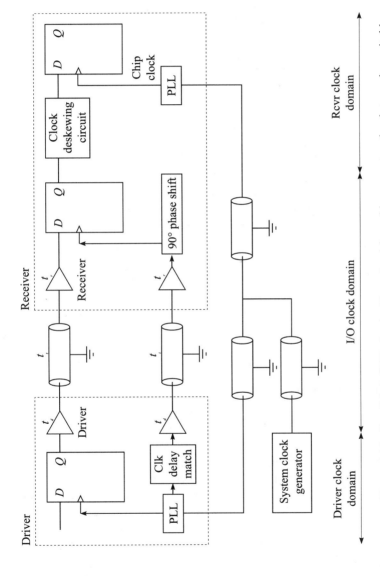

Figure 4-17. Simplified CTC circuit. Here the driver chip generates the data and a suitable strobe pulse. The receiver chip uses the strobe to latch in the data and then uses another circuit to align the external asynchronous data to the internal chip clock.

137

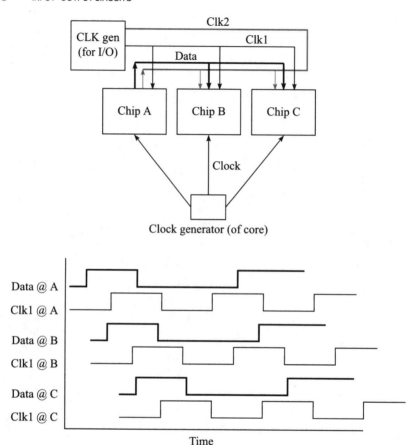

Figure 4-18. RAMBUS I/O scheme. The flight time skew is canceled because the clocks and the data traveling in the same direction have to travel equal lengths. The receiver chip uses the clocks (with appropriate phase shift) to latch the data. The latched data are then aligned back to the internal chip clock.

Even with so many terms (which are present for the common clock scheme too) CTC schemes can have a fairly aggressive performance.

The CTC scheme worked on maintaining tight control on the differential timing between the data and clock, thereby greatly increasing I/O bandwidth. Typically, the CTC schemes may function two to four times faster than a conventional common clock protocol. However, this also implies that the I/O needs to toggle faster, which is dependent on net topology. If, however, faster toggling is not permitted (due to topology or noise issues) but the edge placement is very accurate, then another scheme called phase modulation can be implemented.

4.5.3. Pulse Width Modulation Scheme

Pulse width modulation (PWM) is based on a simple principle of phase modulation. Normally data are sent at each clock pulse, either high or low, and this pattern is repeated as required. In the pulse modulation scheme, the data being sent is first encoded into a pulse width. The received pulse is then decoded using its width at the receiver to recover the data. This scheme is illustrated in Figure 4-19 [Nogami, 1994; Dally, 1996, p. 23; Yamauchi, 1996].

In Figure 4-19 the edge placement capability is τ picoseconds. Thus n such edges will need $n\tau$ picoseconds. In addition, the bus usually will need a minimum time to settle, represented as T picoseconds. This makes the minimum pulse width as T picoseconds and the maximum as $T + n\tau$ picoseconds. The number of data bits that can be represented by the edges is given by

$$n = 2^m - 1$$

$$\mathrm{TR} = \frac{m}{(2^m - 1)\tau + T}$$

$$2^{m_{\mathrm{opt}}}(m_{\mathrm{opt}} \ln 2 - 1) = \frac{T}{\tau} - 1$$

(4-18)

where m is the number of coded bits, m_{opt} is the optimum number of coded bits, and TR is the transfer rate. For example, to code 2 bits, 3 edge positions will be required and the zero coded as no transitions.

The PWM scheme needs to code data at the driver. Usually it takes some additional time to prepare the data and that introduces delays. However, if the pulse width T is large enough, this data encoding can be done after launching the leading edge of the pulse. After calculating the right edge, the lagging pulse edge is sent. The lagging pulse edge can start a new cycle where the falling-to-rising pulse width is measured.

Once the signal is received, it takes some time to decode. Thus the latency of this scheme is at least the minimum of the bus settling time T in picoseconds or the time to encode the data. At the receiver, an additional delay to decode the data is required.

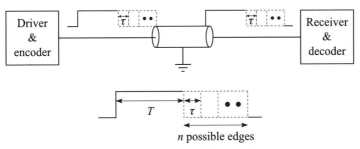

Figure 4-19. Pulse-modulated scheme.

Consider a practical example. Let T be 4 ns and τ be 250 ps. If a normal scheme is used, then for 8 ns two bits of data can be sent over one wire. However, consider the following. In the 8 ns, 16 edges can be coded that represent four bits. This is twice the data sent in a conventional I/O scheme. If T is 16 ns, then in 32 ns two data bits can be sent. If now τ is 112.5 ps whereas this scheme can send eight data bits, a fourfold improvement occurs.

This performance boost comes at the price of increased latency to encode and decode, extra circuitry for coding and decoding, and timing edges to code and decode the information. The major benefit of this scheme is when the available resolution is very fine and the time to settle the bus is relatively large.

Here the common clock method has been discussed in general. Two implementations to increase the bus throughput based on CTC have been illustrated, which made the I/O speed somewhat independent of the system clock. This scheme thereby pushed the number of edges allowed per unit time. In another approach the number of edges was decreased, but by suitable coding the data throughput could be enhanced (over the common clock). In the above discussions the topology was critical as it affects signal integrity and bus settling time. Thus it is worthwhile to examine a few common topologies.

4.6. TOPOLOGY EFFECTS

The topology of the network has a very dramatic effect on the performance of the bus. The topologies may range from many agents (chips) on a single data bus to just two chips communicating. Placing many agents on a bus is an easy an economical way to create a system, although it results in degraded electrical bus performance. The point-to-point communication is usually preferred for very high speed links between two chips.

The terminated point-to-point system in a controlled environment with good termination is the best performance one can obtain electrically. These are usually simpler to construct, and the weakest link in the process is the requirement for connectors (say a mother board connecting to a daughter card). These connectors introduce cross talk, impedance mismatches, layout irregularities, and higher cost [Deutsch, 1994]. So care has to be taken to either eliminate the connector or have a good-quality connector in place.

For high-speed point-to-point communication, system performance can be improved if on-chip termination is utilized. This on-chip termination reduces the bus length so the settling is better, as is the parasitic inductances in the path [Dally, 1996, p. 23].

The high-speed multiagent bus has a lower performance than a point-to-point bus. The main issues, which are illustrated in Figure 4-20, are as follows:

- the increased bus length,
- the stubs connecting the main bus to the chip, and

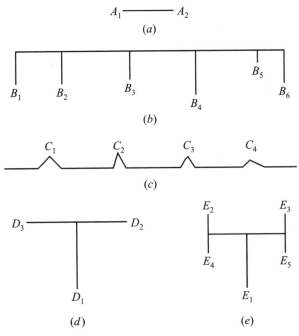

Figure 4-20. Schematic view of some common topologies found in common buses: (a) point to point; (b) a multiagent bus with stubs connecting from the main trunk to the chip; (c) the same multiagent bus with stubs removed by daisy chaining the connection; (d) a T-topology that looks balanced from D1 but is not balanced from D3 and D2; and (e) an H-topology that is balanced from E1 but not from any other agent.

- the uneven environment (i.e., driver in the middle sees a different bus than the ones at the end), which is difficult to optimize for all cases.

The typical design cycle for a multiagent bus is shown in Figure 4-21. Here the first step is to identify the style of the bus required: series or parallel terminated. This is based on the speed, estimated bus length, and number of agents on the bus. Once the termination is fixed, the package and topology must be determined. The primary idea is to minimize the stub lengths. The rise time t_r and fall time t_f of the I/O are then selected based on the requirement that the rise time be greater than at least twice the TOF of the stub (based on similar arguments that the wavelength should be much greater than the stub length). This makes the stub "disappear," and its effect is more of a capacitive loading than a branch in the transmission line. This effect is shown in Figure 4-22 where the effect of the stub is shown at the stub end. Figure 4-23 shows the effect of the stub loading at the far end of a transmission line.

The rise time has to be lowered for the fast case to meet the criterion that it be twice the TOF, but the consequence is that in the slow case the rise time gets

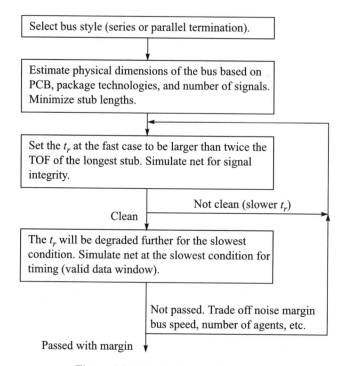

Figure 4-21. Typical bus design cycle.

degraded drastically. Lowering the rise time reduces the available valid data window at the receivers, leading to slower bus performance. In particular the common clock modes are worst affected. Similarly, if there are inductors in the transmission line environment, such as poor connectors, there is a penalty in terms of signal quality degradation. This is shown in Figure 4-24, where the effects of an inductor in the middle of a transmission line are illustrated. The inductor attenuates the high-frequency component from propagating, resulting in a voltage spike at the midpoint and a slower edge rate at the far end.

A similar delay is induced if extra capacitance is introduced in the middle of a transmission line environment. Such capacitance can occur as small stubs, package loading, solder pads, and probe points. The capacitance and transmission line acts as a low-pass filter. Therefore a slow-rising pulse is seen at the point where the capacitor is attached and the far end suffers additional delay. This is illustrated in Figure 4-25.

Another effect of the stubs is that they reduce the bus impedance. The stubs can be thought of as extra load capacitance tapping off a main transmission line and distributing the effect of such capacitance. The bus impedance of such a system is [Intel, 1996] shown in Figure 4-19:

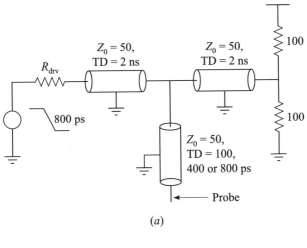

(a)

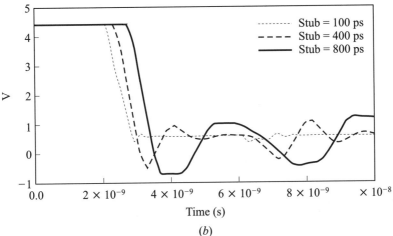

(b)

Figure 4-22. Length of the stub compared to the edge rate determines whether a stub is seen as a branching transmission line discontinuity or can be modeled as a lumped element at the junction. The longer stubs cause degradation in signal quality. The effect of the short stub (100 ps) is negligible whereas the transmission line with a longer stub (800 ps) rings before settling.

$$Z_0' = \sqrt{\frac{L'}{C'}} = \sqrt{\frac{L_0}{C_0 + C_d}} = \sqrt{\frac{L_0}{C_0\left(1 + \frac{C_d}{C_0}\right)}} = \frac{Z_0}{\sqrt{\left(1 + \frac{C_d}{C_0}\right)}} \qquad (4\text{-}19)$$

where C_0 is the main trunk total capacitance (capacitance per unit length × length) and C_d is the total stub (capacitance per unit length × length), device, and socket capacitances of all the attachments to the main transmission line.

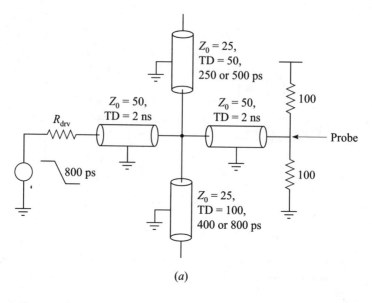

(a)

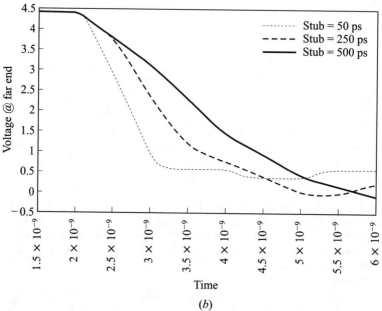

Time

(b)

Figure 4-23. Far-end received signal is a function of the stub length introduced at the midpoint. The longer the stub, the higher the loading at the midpoint, thereby delaying the signal longer.

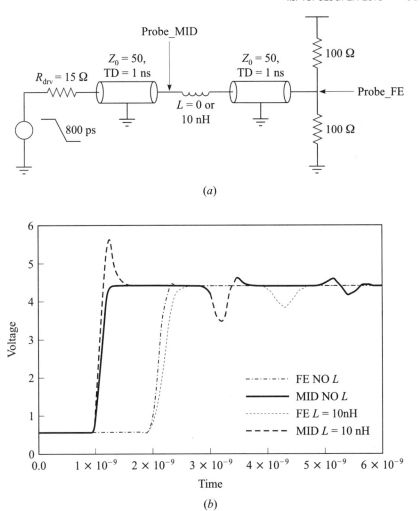

Figure 4-24. Inductor in the transmission line causes a sharp overshoot. It also acts as a filter in reducing the high-frequency content of the signal, which degrades the transmitted waveform. The above waveforms show the effect of placing an inductor at the midpoint of a transmission line. There is a sharp overshoot at the midpoint when an inductor is placed. This spike is due to the $L \, di/dt$, which is filtered. At the far end this filtering delays the arriving signal.

As shown here, the stub length forces a slower edge rate on the signals and the stubs cause delays. In addition, connectors, probe points, and so on, all delay the signal propagation. The increased delays and the faster buses may cause situations where the bus does not settle completely in a data window. If the topology chosen is one where the bus does not settle completely before the next data bit is initiated, then the previous data bit affects the current data bit.

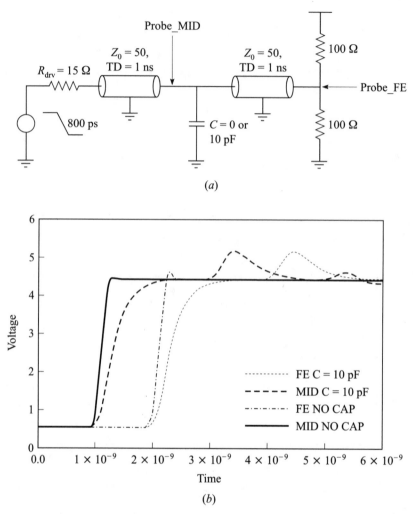

Figure 4-25. Effect of a capacitor in the transmission line environment is a lowering of the impedance at the point of attachment. The capacitor filters out the high-frequency content of the step, thereby lowering the rise time at the far end.

This induces jitter in the data, which reduces the available window at the receiver. For example, consider the circuit in Figure 4-26, showing a transmission line of impedance 50 Ω and delay 1.2 ns driven by a 70-Ω driver. The data bit cell period is 2 ns, and clearly the line does not settle before the next bit cell is launched. Thus at high frequency (HF), toggling the bus does not settle. At half the frequency (LF), the bus is close to settling. The data pattern dependency introduces a timing jitter than can be seen when the two falling edges are compared, as shown in the Figure 4-26. This jitter can be seen at the same

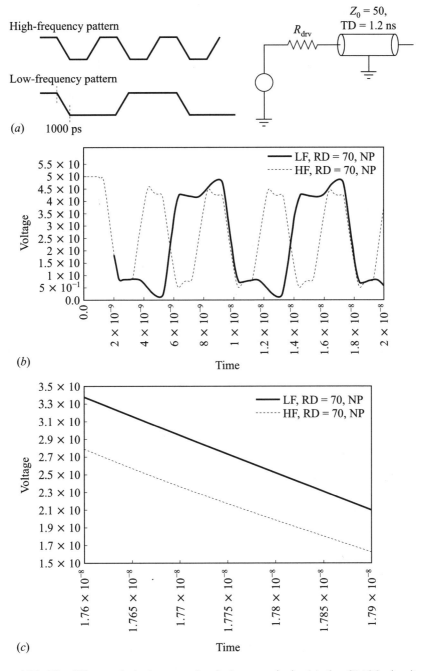

Figure 4-26. The ISI negatively impacts the timing margin in (a) the CMOS circuit shown with a weak driver having a slow edge rate similar to that found in the slow condition. The skew between the low toggle and high toggle rate data is shown in the falling edges around 17 ns and (b) a blown-up view. The ISI in this case is 140 ps for a 250-MHz signal. For a CTC scheme, the setup and hold window are each 1 ns; therefore this jitter represents a 14% timing impact.

receiver at different times (if, say, the data pattern is 1010100110011) or between different receivers on a bus receiving different patterns (i.e., say a high- and low-frequency toggling pattern). If this network is used in a CTC scheme, the jitter introduces a 14% penalty in the setup and hold window (i.e., a 140-ps jitter in a 2000-ps bit cell of which half is allocated to setup and half to hold).

The ISI can be reduced by having a well-matched driver to the transmission line and a fast edge rate.

Until this point we have been discussing a unidirectional I/O approach, that is, one driver and multiple receivers. In a bid to push the throughput per wire, it is possible to have two-way transmission. Here two drivers can drive a net and simultaneously receive. Such bidirectional schemes will be examined next.

4.7. BIDIRECTIONAL SIGNALING

In the previous sections only unidirectional signaling circuits have been discussed. However, it is possible to send bidirectional signals on a transmission line under very specific circumstances. Consider the scheme shown in Figure 4-27. It shows a method to send data in both directions over one wire simultaneously. This can be viewed as dividing the complete amplitude into two halves to allow the two devices to talk on the same bus simultaneously. By trading circuit complexity, the data rate is doubled over one wire [Mooney, 1995; Takahashi, 1995].

The core idea is that if a wave is launched in a transmission line and the transmission line is perfectly terminated, no reflections come back to the source. So waves initiated by A are terminated by driver B, and vice versa. However, consider the voltage levels in this scheme. When both drivers are driving whether high or low, the line voltage is correspondingly high or low. However, when the drivers are driving opposite-polarity signals, the line voltage is half of V_{cc}.

When the line is at either V_{cc} or V_{ss} the corresponding value of the incoming data is clear to both receivers. However, the natural question is how to decode the incoming value when the line is at half V_{cc}. For this decoding, the state of the outgoing data at the driver is required. If the outgoing bit is high, then the line values can be either half of V_{cc} (i.e., if the opposite end drives a low) or V_{cc} (i.e., if the opposite end drives a high). To make this comparison, a differential amplifier receiver is required. If the reference voltage selected is three-quarters of V_{cc}, the differential amplifier can then distinguish whether the incoming data are a high or a low. This logic is presented in Table 4-5.

Since the voltage swing is small (half of V_{cc}), it is critical to compare it against a good reference voltage. Any errors in the reference voltage will directly diminish the noise margin. new method to generate and distribute the reference voltage is shown in Figure 4-28 [Mooney, 1995]. In this scheme the reference voltages are generated at the driver and receiver chip and then

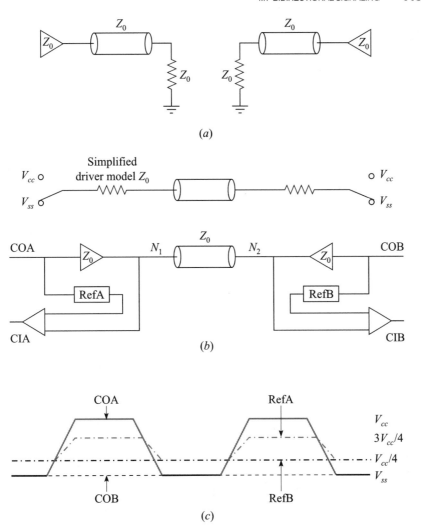

Figure 4-27. (*a*) Conceptual view of bidirectional signaling scheme by superpositioning two unidirectional point-to-point schemes. (*b*) A bidirectional scheme is realized. (*c*) The reference voltage of the receiving differential amplifier must be changed as a function of the state of the driver. To minimize ISI and reflections, the bidirectional scheme requires an extremely clean termination.

combined to average them. This scheme allows some common mode noise (with the signals) to be coupled into the reference and provides an AC reference that is better than a purely on-chip-generated reference.

An interesting point to note is that since the line is perfectly terminated no reflections are generated as long as transmission lines can be used. For high-speed I/O, fast edge rates are required; therefore, the total resistance of the

TABLE 4-5. Relationship Between Outgoing Data, Reference Selected, and Line Voltage

COA	COB	Reference A	Reference B	Line
0	0	$\frac{1}{4}V_{cc}$	$\frac{1}{4}V_{cc}$	V_{ss}
0	1	$\frac{1}{4}V_{cc}$	$\frac{3}{4}V_{cc}$	Half v_{cc}
1	0	$\frac{3}{4}V_{cc}$	$\frac{1}{4}V_{cc}$	Half V_{cc}
1	1	$\frac{3}{4}V_{cc}$	$\frac{3}{4}V_{cc}$	V_{cc}

transmission line should be much less than the characteristic impedance to avoid *RC* behavior.

The above scheme works only in a point-to-point environment. Also it requires a very uniform system and good termination. Any deviation from these dramatically lowers the transfer rate.

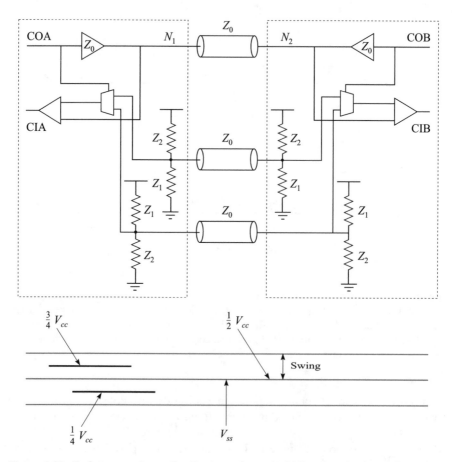

Figure 4-28. Reference voltage distribution scheme in bidirectional scheme [Mooney, 1995].

As is clear in this discussion, a key ingredient is that the source and termination values must be very close to Z_0 over all possible temperatures, voltages, and processes. If this is not maintained, both the timing and noise margins rapidly diminish. The only method to ensure that the output resistance is constant is to compensate the I/O buffer over all possible ranges of process, voltage, and temperature (PVT).

4.8. COMPENSATION SCHEMES

During fabrication, as in any manufacturing step, there are target specifications and there are manufacturing tolerances around each specification. For example, an oxide thickness specification translates to slower devices for thicker oxides and faster devices for thinner oxides. Similarly, through the myriad of steps like photolithography, ion implantation, doping, metallization, and oxides, all contribute to various aspects of the final circuit. In Figure 4-29, the cumulative effect between a device that is "slow" and one that is "fast" is shown in terms of $I–V$ characteristics in an NMOS device. The large ratio (> 2) between driver currents is an effect of mainly the threshold voltage differences and applied voltage for the fast and slow cases (both V_{gs} and V_{ds}) [Dally, 1996, p. 5]. If such a device was used as a driver element, large variations in driver strengths and slew rates (from the predriver) should be expected. This in turn affects the timing of the outgoing signals.

With higher speed I/O, implying shrinking timing budgets, such variations cause a decreased operational window for common clock schemes and a larger latency (time before valid data are available at the receiver core logic) penalty for the CTC operations. Any decrease in the fast and slow skews can usually be translated into better timing budgets and signal integrity, resulting in improvements in margins or the system I/O speed.

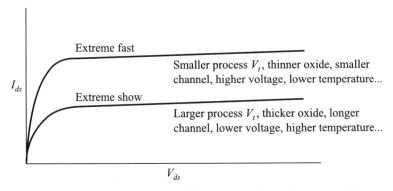

Figure 4-29. Current spread in NMOS device in fast and slow conditions.

Balancing the requirement of setup and hold time limits system speed. One method to improve system speed is to provide circuit compensation. The compensation allows the designer to speed up the slower I/O driver and receiver speeds and increase the driver strength for the slow part. At the same time the fast part is slowed down to match the slow part in speed and drive strength. In an uncompensated case, speeding up the slow case will in turn speed up the fast case and may counter the gains in the slow case. Also there may be significant differences between fast and slow cases in properties such as driver strength, clock-to-output delay, and driver slew rates. Thus, benefits of compensation are less skewed timings, improved duty cycle, better signal integrity, reduced power consumption, lower *di/dt* noise, and improved overall system speed.

Compensation schemes for I/O circuits can be broadly classified into slew rate compensation and output resistance compensation. These can be implemented in various methods using digital or analog schemes. The spectrum of schemes is shown in Figure 4-30.

Traditional compensation schemes have not been favored because I/O was relatively slow and the timing margins were not so aggressive. Additionally, the compensation schemes consumed extra area and needed extra testing so for slower systems these were not cost effective. However, with increasing system speeds, compensated I/O is becoming essential. In any compensation schemes there are three tasks to be completed:

- compensation generation,
- distribution, and
- usage.

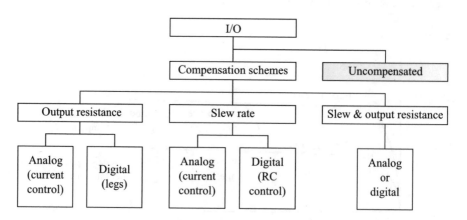

Figure 4-30. Compensation schemes can be realized as analog or digital circuits and can be used to control the slew rate and the output resistance.

These will also bring the inherent issues in compensation to the forefront:

- noise sensitivity, board to chip;
- update of digital values;
- locality of generation and different-use point;
- noise sensitivity—on chip; and
- noise sensitivity—reference generation.

These tasks are crucial for the correct implementation of a compensation scheme and will be considered next.

4.9. ANALOG COMPENSATION

A number of analog compensation styles are possible and one such scheme is shown in Figure 4-31. The compensating device (NMOS or PMOS) is compared to a known resistor (programmable) and the voltage is fed back to an operational amplifier (opamp). The opamp compares this voltage to that of a known reference voltage. Thus, depending on the chip processing (fast or slow), the correct voltage is generated to make the device drains match the reference voltage. Using this compensation voltage, the I/O buffers can be biased. The strengths of drivers and predrivers can be adjusted by ratioing the gate widths with respect to the compensated N and P devices.

Before the compensating voltage can be used, it has to be distributed on the chip to each buffer. Since the distribution interconnection can couple noise from other digital signal lines and be lengthy (high R, so slower to discharge the injected noise), it should be closely shielded. An effective shield is the addition of power interconnects in parallel with the compensation voltage interconnect. This shields the sensitive wire in two ways: by spacing other wires to nonminimum space and by providing capacitance to the power rail. This additional capacitance allows charge sharing of the injected noise voltage with the power supply node, thereby attenuating the effect of the injected noise spike. Figure 4-32 shows this concept. Utilizing such a scheme one compensation source can supply multiple buffers.

Other distribution schemes employ a current distribution method. In this method, a reference current is generated based on the compensation and then sent to the point of use. These are shown in Figure 4-33. Here a current is generated using a current source MP_1 that is a function of the analog bias voltage. The generated current is again converted into a voltage using the NMOS MN_1. The n_{bias} voltage can then be used once again to generate p_{bias} compensation voltage. Usually, one global current source will source one local current generator. For a current distribution scheme the reader is referred to Satyanarayana [1995; Allen, 1987].

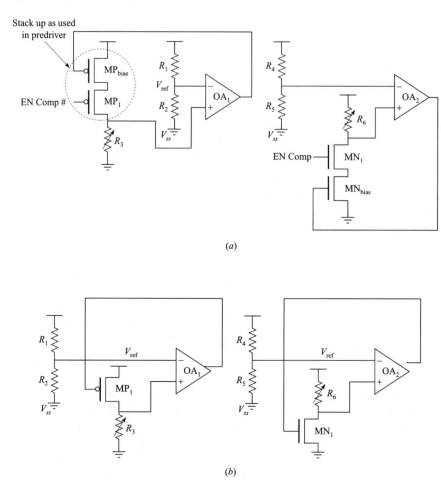

(a)

(b)

Figure 4-31. Analog bias generation scheme that can be used to compensate I/O buffers. (a) Transistor stack reflects the stack up in a predriver. (b) This option is convenient as it reflects a normal output buffer and no edits are required as in (a). These schemes are based on equalizing the voltage drop across a resistor and transistor, signifying the transistor resistance (total voltage/total current) is equal to the programming resistance.

It is worthwhile to contrast the current and voltage distribution schemes. The voltage distribution method is a low-output-resistance voltage source system, whereas the current distribution system is based on a current source (high source impedance). When equal noise is injected into this system, the voltage distribution system behaves better as it is able to source the required offset current to cancel the injected noise current swiftly. For the current distribution system, the same injected current (ΔI) needs ($\Delta I / I_{bias}$) time to settle to the DC

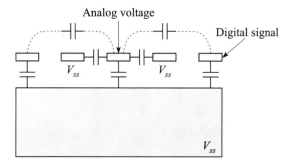

Figure 4-32. Analog voltage delivery scheme using V_{ss} wires to shield and charge share the injected noise. The digital signals should not run parallel and close to the analog wires over long lengths.

value. Thus for quick settling, I_{bias} has to be large or the noise coupling has to be minimized.

Once the shielded analog compensation signal is distributed to the local buffer area using either the current or voltage distribution schemes, it can be used to bias the predriver transistors, as shown in Figure 4-34. This predriver is a current-starved inverter. In this region of operation it supplies a constant current. Once V_{ds} becomes small ($|V_{ds}|$, $|V_{gs} - V_t|$), the linear operation occurs, and the constant current operation no longer holds. Consider the fast case.

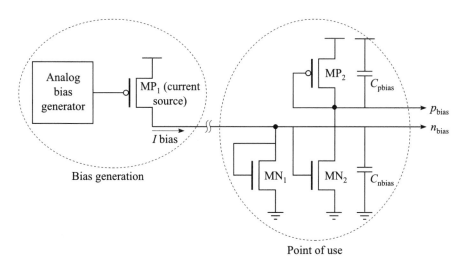

Figure 4-33. Simple current distribution scheme using current source MP_1 to source a current based on the analog voltage and a current pick-up MN_1 to sense this current at the local receiving end. Using the current mirror technique, a single bias voltage can be split into two.

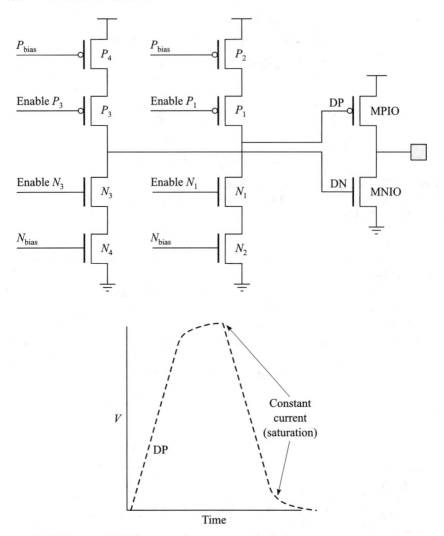

Figure 4-34. Simple analog slew rate control circuit. Transistors N_3 and P_3 are the digital switches and the transistors N_4 and P_4 control the actual slew rate to the driver gates [Senthinathan, 1994, p. 104].

Here a low V_{gs} is sufficient to provide a match between the compensation resistor and the NMOS transistor. This low V_{gs} is communicated back onto the predriver NMOS. Since V_{gs} is small, the predriver stays in a constant current mode for a longer duration. This limits the slew rate (The RC slew rate being faster). For the slow case, the NMOS compensation gate voltage is higher and the slew rate is limited. Thus for a slow case, the transistor has more linear operation than the fast case.

The preceding paragraphs described a compensation generation scheme based on voltage divided between a programmed resistor and a transistor. There are other current reference circuits available. One is shown in Figure 4-35. It consists of a PMOS current source P_1 and P_2 and two biasing transistors N_1 and N_2. The bias transistors provide equal current to both the transistors N_1 and N_2. Since there is a constant resistor in the current path for transistor N_1, it acts as the feedback. If the current increases in N_1, the voltage drop $V_{gs}(n_1)$ decreases, so the current is reduced.

The current in the diode scheme is given by [Alvarez, 1995]

$$I = \frac{\sqrt{K} V_{TH} \ln[A(d_1)/A(d_2)]}{\sqrt{L/W_{n2}} - \sqrt{L/W_{n1}}} \qquad (4\text{-}20)$$

where L is the channel length (kept long), W_{n1} and W_{n2} are the transistor widths, and K is the product of mobility and half of the gate oxide capacitance per unit area $(C_{ox}/2)$. It should be noted that W_{n1} must be larger than W_{n2} for this scheme to function.

In the resistor-biased case, the current and voltage are related by [Ooishi, 1995]

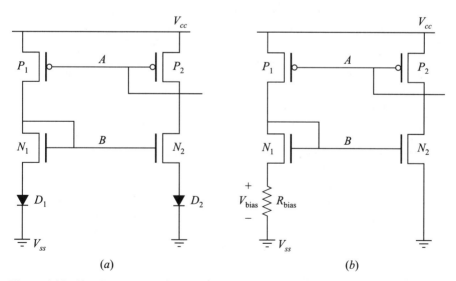

(a)   (b)

Figure 4-35. Simple current reference circuits: (*a*) a diode-based partial current reference [Alvarez, 1995, p. 631] with no external components; (*b*) current reference based on a resistor [Ooishi, 1995]. The ratio of the transistors sets the bias voltage. The device N_1 must be larger than N_2 for this scheme to work.

$$V_{\text{bias}} = \left(\frac{kT}{q}\right) \ln\left(\frac{W_{n2}}{W_{n1}} \frac{W_{p1}}{W_{p2}}\right)$$
$$I_{\text{bias}} = \frac{V_{\text{bias}}}{R_{\text{bias}}}$$

(4-21)

Provided there is sufficient feedback from the resistor, the current can be compensated fairly easily in this scheme. However, the resistance should be kept small such that the RC time constant is much smaller than the highest noise coupling frequency with a significant energy component.

4.10. DIGITAL COMPENSATION

The analog schemes are sensitive to noise (as all other analog schemes). This is true in the generation of the compensation voltage and its distribution.

An option is to use digital compensation schemes, and one method is shown in Figure 4-36. Here a circuit similar to the analog case may be used to generate the compensation factor (a series of bits). The calibration transistor is broken into sections. Each leg is then controlled by a control bit. All the "on" legs together represent the driver strength. The control bits are derived from a counter that is fed by a comparator. The comparator senses the voltage division between the resistor and the transistor legs and compares it to a reference voltage [Gabara, 1992; Sekugichi, 1995; DeHon, 1993; Takahashi, 1995, p. 40; Pilo, 1996]. A key difference between analog and digital compensation is that in the digital scheme when a leg is turned on it is fully on (V_{cc} at the gate of an NMOS device), unlike the analog case, where all the legs are partially on.

The digital comparator can change states as the feedback loop time permits. Usually the state (increments/decrements) allowed per clock is small (say 1) and the updating clock is relatively slow (microseconds). In this fashion the compensation is altered slowly and the bus is not jittered wildly due to multiple compensation bits switching and updating simultaneously. The distribution of digital compensated signals is easier because they are all at normal CMOS voltage levels and not at an intermediate analog level, and thus they are less noise sensitive. No special precautions have to be taken, except those similar to the other CMOS signals. The digital bits can be latched into a local latch and distributed from there to the individual buffers. This in turn allows a serial compensation bit transmission from a bias generator to the local latch, which after capturing the serial data provides it in parallel to the I/O buffer. One such implementation is shown in Figure 4-37.

Since the digital scheme has discrete jumps as the driver strength of the transistors varies discretely, the jumps should be kept small such that no functionality is affected. Two methods can be employed to minimize these jumps (Figure 4-37). One is to have very fine resolution jumps, maybe several times the noise level. If the linear compensation is used, then the number of legs and

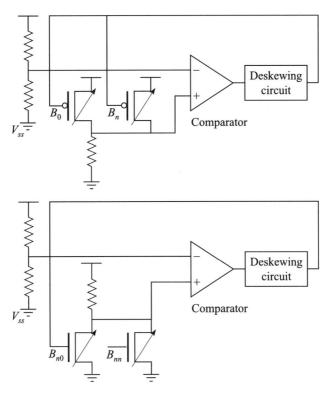

Figure 4-36. Digital bias generation scheme. It may be similar to an analog scheme except that the transistors are not one device but a number of parallel bits capable of being switched on and off independently.

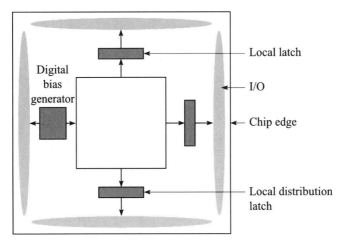

Figure 4-37. Schematic of digital compensation scheme showing generation and distribution method.

compensating bits will be large. The number of discrete levels is $n - 1$, where n is the number of legs. The other option is to use binary weighted legs [DeHon, 1993]. Here, the jumps and the number of control bits are minimized by providing $2^n - 1$ levels. However, this option can lead to complexity in the layout of the I/O driver. Yet another method is to have approximate equal percentage change for each step (addition or deletion). With reference to Figure 4-38c, the ratio of legs that are needed is given by

$$\frac{1+x}{1} = \frac{1+x+y}{1+x} \leq \text{Max\_Allowable\_\%\_Change} \qquad (4\text{-}22)$$

In the previous schemes a compensating resistor has been used to adjust the driver strength and its slew rate. Consider a resistor specification that is tuned for a 65-Ω transmission line during system design and specification. However,

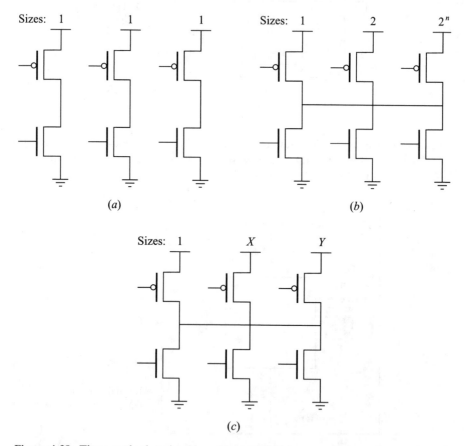

Figure 4-38. Three methods to implement legs of digital compensation scheme using (a) uniform legs, (b) binary weighted legs, and (c) equal percentage change.

during manufacturing, the transmission line can deviate from the specification (say 5 or 10%). In this case the driver is no longer accurately matched to the system transmission line impedance, with corresponding negative effects. Also it should be noted that the resistors themselves have some tolerance, thus increasing the possible mismatch between the driver and the transmission line. For these reasons, a compensation based on matching the transmission line and the driver that is adaptable to the specific board's environment is desirable [DeHon, 1993]. Such schemes utilize the transmission line itself as the calibration impedance. In this technique a step is driven out onto the transmission line. If the high-going step is applied and the driver is perfectly matched, then the resulting voltage level at the pad will be $\frac{1}{2} V_{cc}$. If the step is lower, it indicates that the driver has higher resistance than the transmission line impedance. Therefore the drive strength should be increased. If the step height is higher than $\frac{1}{2} V_{cc}$, then the driver impedance should be lowered. The voltage level is sensed by the comparator, and by suitably applying a feedback, correct setting and updating can be arranged. Also the ramp rate should be sufficiently large that the parasitic inductance and capacitance at the pad do not interfere with the measurement. Similarly the sampling point should be selected when these parasitic effects do not interfere with the measurement process.

4.11. FREQUENCY-BASED COMPENSATION

There are weak spots in both the analog and digital schemes:

- External pins provide a path for the calibration resistor and their noisy power supply termination.
- Bias generation is in one spot and distribution is to many local points of use, each with its own different microenvironment.
- Only the resistor value is calibrated and the capacitance variations are not fully compensated. Thus the slew rate compensations only provide the resistive compensation (and some capacitive compensation as C_{ox} is linked to I_{ds}).

A scheme that can avoid these complications is based on frequency [Shirotori, 1991]. Whereas resistors have tolerances ranging from 1 to 5%, clock sources have stability and accuracy measured in parts per million. Thus a crystal that is easily available can be an excellent standard reference for a compensation scheme. The frequency is converted into an appropriate voltage by the use of a suitable phased-locked loop (PLL) or a delay-locked loop (DLL). The PLL is hard to implement and considerably larger. A DLL is usually smaller and can provide the same functionality.

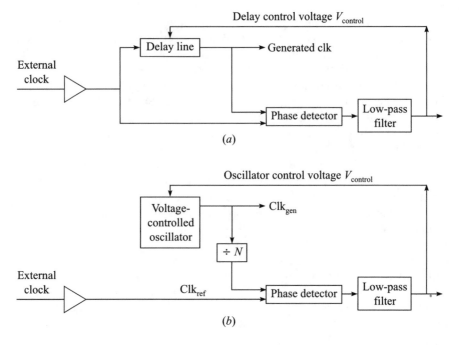

Figure 4-39. Schematics of (a) delay-locked loop and (b) phase-locked loop. Both digital and analog PLL/DLL can be used to provide respective compensation styles. Differential oscillators and delay lines provide lower timing jitter compared to single ended implementations.

The PLL is a device based on establishing a fixed phase between a reference clock and an internal voltage-controlled oscillator (VCO) clock. This is illustrated in Figure 4-39. The VCO needs higher voltages at slower components, voltage, and temperature conditions, than at the faster component and condition. This voltage can be used as a bias voltage for compensating the I/O. One major factor that compensation helps to alleviate is the noise generated when a large number of I/O buffers switch. (See Figure 4-40.) The next section discusses the significance of the noise generated and methods to control it.

4.12. SIMULTANEOUS SWITCHING OUTPUT NOISE

A very important factor affecting the output driver performance is the di/dt noise injected into the peripheral V_{ss} or V_{cc} due to a number of simultaneous switching outputs (SSOs). This effect is shown in Figure 4-41. In the following example, the main current is assumed to be the crowbar current, which is the short circuit current during the transition time when both the NMOS and

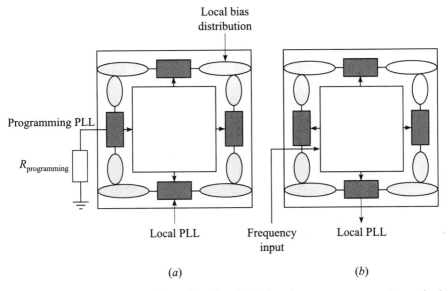

Figure 4-40. Compensation scheme based on PLL signal recovery to generate required bias signals for local area on chip.

PMOS are partially on. The noise generated by the $L\,di/dt$ term is shown in Eq. 4-23. the noise biases the source positive, thus reducing V_{gs} on the driver, which reduces current flow. Therefore, as more and more buffers switch simultaneously, larger di/dt noise is generated. However, the noise does not grow

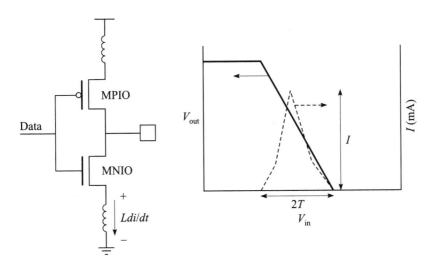

Figure 4-41. The I/O buffer and the currents used in the analysis [Senthinathan, 1994, p. 24]

linearly but has a square-root dependence because of the reduced gate-to-source drive due to noise (for details see, Senthinathan [1994, p. 24]).

Based on the crowbar current, the ground bounce of the simple system shown in Figure 4-41 can be calculated to be [Senthinathan, 1994, p. 24]

$$V_n = V_k + \frac{T}{L} \frac{p}{nK} \left[1 - \sqrt{2V_k \frac{Ln\,K}{pT}} \right] \qquad (4\text{-}23)$$

where V_n is noise voltage, T is current rise time, p is the number of parallel paths (pins), n is the number of I/O buffers switching, K is a device constant, $V_k = V_g - V_{tn}$, and V_{tn} is the threshold voltage of the NMOS device.

In an actual I/O buffer, designers take great care to minimize the crowbar current. This is done by skewing the predriver circuit. When the NMOS device is turned on, the PMOS driver is turned off faster, and similarly when the PMOS is turned on, the NMOS driver is turned off faster. This is illustrated in Figure 4-42. In reality, there is only a limited time allowed for switching on and off the drivers, so in an aggressive design there is always some crowbar current.

When the driver is utilized on a PCB, the major portion of the current drawn is for charging or discharging a transmission line. Thus Figure 4-43 shows a real-life I/O buffer setup and a current waveform. The ground bounce observed for such as system is shown in Figure 4-44. As the number of drivers increases, the ground bounce increases but does not grow linearly. This is again due to the reduced gate-to-source voltage caused by the SSO noise. Also, as the number of pins is increased, the ground bounce is reduced and becomes increasingly linear (in the range shown). This linearity should be expected because the noise voltage is reduced; thus the gate-to-source voltage for each driver is not reduced.

Often it is desirable to model and simulate this SSO noise. This can be done by utilizing a capacitor or a transmission line being switched by a buffer. This is shown in Figure 4-45. For the capacitor model, if the buffer output resistance is assumed to be R_0, then the peak current in the circuit is V/R and the rate of current decay is V/RC. Consider the transmission line model. When a driver of output impedance $R_0 = Z_0$ pulls the transmission line low, then the peak current is $V/2R$, or half the current of the buffer capacitance model.

If the RC model is used, the SSO noise is extremely overestimated. To compensate for this noise, a larger driver will be used. In reality there may be no need to strengthen the drivers as the di/dt noise is lower than simulated in the SSO model. In fact, the overdrive may cause larger reflections at the receiver and can cause latchup or oxide reliability issues [Hu, 1994].

There may be cases where the drain current is limited by the saturated velocity of the electrons and holes in the device. In such velocity-saturated cases I_{ds} is severely reduced [Pierret, 1990]. The I_{ds} is now given by Eq. 4-24

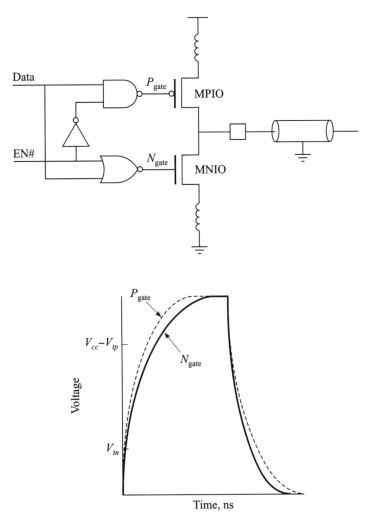

Figure 4-42. Predriver skewing to reduce crowbar current and have reasonable delay time through buffer.

and the ground bounce in such a case is obtained by differentiating the $I_{d,\text{sat}}$ with respect to time [Yang, 1996]:

$$I_{d,\text{sat}} = \beta_o V_c (V_{gs} - V_t - V_c)$$

$$V_g(t) + \frac{dV_g(t)}{dt} = \frac{L_g n \beta_o V_c V_{dd}}{2t_r}$$

$$\beta_o = \mu C_{ox} \frac{W}{L_{\text{eff}}}$$

$$V_c = \frac{v_{\text{max}} L_{\text{eff}}}{\mu}$$

(4-24)

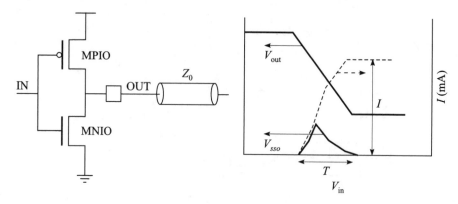

Figure 4-43. Fast I/O drivers see a transmission line at the pad.

where V_g is the ground bounce voltage, V_t is the threshold voltage, V_c is the critical voltage, and t_r is the rise or fall time at the input to the output device. Using Laplace transforms, V_g can be written as

$$V_g(s) = \frac{1}{1 + Ks} \qquad (4\text{-}25)$$

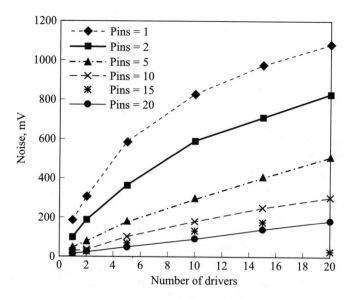

Figure 4-44. Ground bounce of NMOS drivers discharging a transmission line ($Z_0 = 65\,\Omega$). The larger the number of available pins, the lower the noise and the more noise voltage scales linearly with the number of drivers.

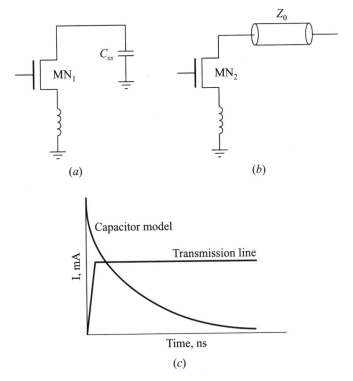

Figure 4-45. An SSO simulation setup: (*a*) load modeled as a capacitive load, (*b*) load modeled as a transmission line, and (*c*) expected currents.

where $k = L_g n \beta_0 V_c V_{dd} / 2 t_r$ and then finding the Laplace inverse. This gives the gate voltage as

$$V_g(t) = \frac{L_g n \beta_0 V_c V_{dd}}{2 t_r} \left(1 - e^{(t - t_t)/L_g n \beta_0 V_c} \right)$$

$$t_t = t_r \frac{V_t}{V_{dd}}$$

(4-26)

For $t_r - t_t > 3$, the exponential term is small ($<5\%$) so the equation can be simplified to

$$V_{g,\text{max}} = \frac{L_g n \beta_0 V_c V_{dd}}{2 t_r}$$

(4-27)

Equation 4-27 can be used to trade off the driver size, the number of buffers toggling, the edge rage, and the number of wire bonds (actually the inductive path) required to meet the required ground bounce.

In this section, causes of SSO have been examined. Some recourses to the SSO are to slow down the edge rate for which buffer compensation is very useful. Another technique may be required if a compensation scheme is not available.

4.12.1. Design for SSO Reduction

4.12.1.1. Predriver Skewing   A method to reduce the SSO is to prevent the complete buffer from switching on simultaneously. This can be partly done by skewing the predriver turn on. Additional skewing in between legs can also be implemented. This can be achieved by making an *RC* delay line using the *C* of the gate and *R* (poly or diffusion), as shown in Figure 4-46. Another version is to use the gate itself as an *RC* line (Figure 4-47). Usually with a silicided process, the gate resistance is small, and this may not be an

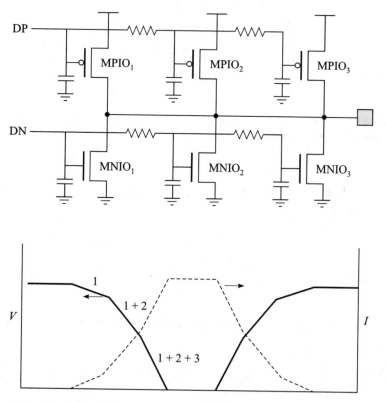

Figure 4-46. Gate driver for controlled slew rate and switching characteristics of the buffer [Senthinathan, 1994, p. 113; Annaratone, 1986, p. 270]. In this buffer, since both the turn-on and turn-off are gradual, some overlap will occur when both the NMOS and PMOS are both partially on.

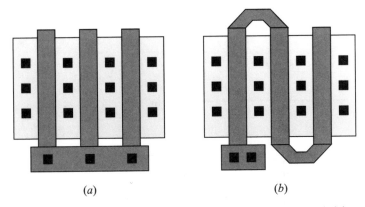

(a) (b)

Figure 4-47. Implementation of resistive slew rate control: (a) normal driver with all gates tied in parallel; (b) slew rate controlled version where gate is tied in series [Annaratone, 1986, p. 270]. With low-resistivity silicided gates, this approach cannot provide good slew rate control unless sufficient snaking of the gate is implemented.

option. If the poly or diffusion has been silicided and therefore resulting in an ineffective resistor, pass transistors can be used to replace the resistors. One such scheme is shown in Figure 4-48. This allows the legs to be turned on with some delay with respect to each other and reduce di/dt.

The previous slew rate control schemes were implemented by either the RC product or the resistance. Alternatively, C can be adjusted to provide a constant RC slew to the output driver, as shown in Figure 4-49. In this case the bits to control the capacitive loading are provided by a digital compensation scheme.

The implementation attempts to control the rate of the turn-on of the driver transistors, lowering the SSO noise. Another approach is to have better control of the driver by using an inverted NP stack-up, thereby reducing the SSO noise.

4.12.1.2. NP-Inverted Stack Driver
It has been observed that even with slew rate control of the predriver the output slew rate is not effectively controlled. The problem can be traced to the turn-on characteristics of the MOS devices. This configuration acts as a common source amplifier [Millman, 1987, p. 136], and as soon as the transistor is biased ($V_{gs} > V_{tn}$), the change in output voltage gains rapidly as small changes in input voltage.

To address this issue, a source follower [Millman, 1987, p. 429] circuit is required, shown in Figure 4-50. Here the regular CMOS circuit has been inverted. Now if a slew rate controlling a predriver is used, the pad voltage will follow the gate voltage less a threshold voltage. This circuit, however, has a reduced voltage swing between a high of $V_{cc} - V_{tp}$ and a low of V_{tn} [Hanafi, 1992]. This reduced voltage swing reduces the current drive required, which in turn reduces di/dt. In addition, the current tailing as the devices move toward cut-off (i.e. V_{tp} for the pull-down PMOS and $V_{cc} - V_{tn}$ for the NMOS) leads to

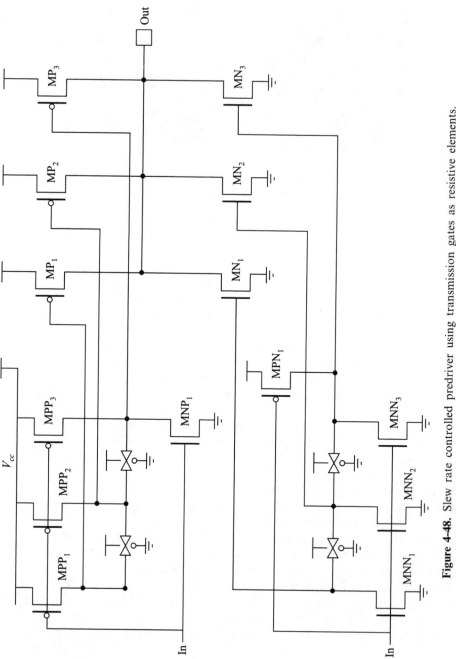

Figure 4-48. Slew rate controlled predriver using transmission gates as resistive elements. Transistors MPN_1 and MNP_1 are turn-on controls and MN_1 and MN_2 and HPP_1–HPP_3 are for fast turn off.

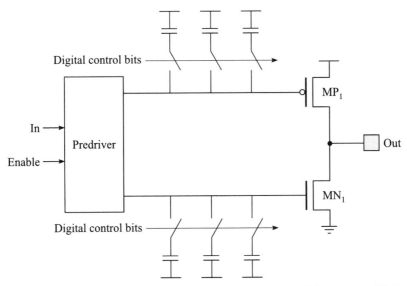

Figure 4-49. Slew rate adjustment by controlling loading on predriver output [Gabara, 1997].

further lowering of di/dt. A word of caution when using schemes terminated in-series: The device cutoff can also lead to poor termination by the driver.

To restore the swing to full CMOS levels, a normally configured CMOS stage but one that is much weaker in size can be added to the driver. Since this scheme can control slewing at the output driver, di/dt is reduced. However, the

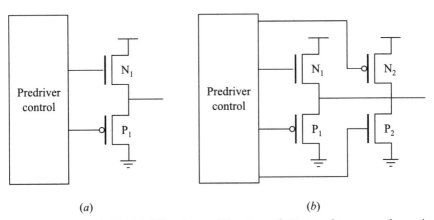

Figure 4-50. Inverted CMOS I/O scheme. The N_1 and P_1 transistors are the main drivers and are configured in the source follower mode [Hanafi, 1992]. Option (a) is not rail to rail, but by placing N_2 and P_2 the full-rail option can be restored.

drivers consist of "wrongly" placed N and P devices and their sizes can be prohibitively large.

4.12.1.3. Differential Signaling   The SSO issue is acute when all the buffers simultaneously swing from one state to another. This generates the maximum $L\,di/dt$ noise. Methods to control noise with predriver slew rate and driver impedance control have been discussed. Another approach to reduce SSO effects is to use differential signaling. In this scheme every bit has its complement transmitted. Therefore the total switching current is theoretically zero. This is illustrated in Figure 4-51b, where a simple CMOS buffer has been configured in a differential mode. The current is simultaneously sunk and sourced from different supplies that are generally well coupled using decoupling capacitors. This effectively cancels out the AC currents; therefore, the ground bounce is not experienced. However, any nonsymmetries in the power supply paths will show up as deviations from the exact complementary behavior. Despite these limitations, this scheme serves well as a differential scheme. For further refinements current switched methods based on matched devices and transmission lines can be used [Rainal, 1994; Dally, 1996, p. 14].

The differential scheme has other advantages. If the signal and its complement are closely routed, then the cross-talk and ground bounce effects are also reduced. At the receiver, the differential input creates twice the input ramp rate, ensuring swift and clean transitions.

4.12.1.4. SSO Reduction Using Packaging Options   The cause of SSO generation was the large instantaneous current flowing through the package inductance. The silicon attempted to control the instantaneous current de-

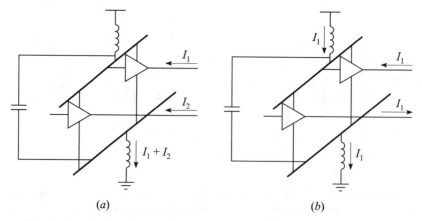

Figure 4-51. (*a*) Single-ended signaling scheme and (*b*) differential scheme.

mand. The other option is to reduce the package inductance. this can be done using several newer packaging technologies:

- Controlled collapse chip connection (C4) as opposed to wire bonds to connect the die to the package. This reduces the inductance from 2–4 nH per wire bond to 200–400 pH per C4 bump. This primarily helps the core and I/O power supply distribution.
- Ball grid array (BGA) package as opposed to pin grid array (PGA) mounting to the PCB. This can help reduce inductance from 2–3 nH per PGA pin to under 500 pH per ball.
- Land grid array (LGA) connections between the package and the PCB can keep the inductance as low as 500 pH.

By trading off the electrical performance, reliability, cost, and other system requirements, the right package can be selected that will reduce the overall inductance and therefore the ground bounce. The packaging technology is making rapid progress, and it is well worth the I/O designer's time to keep abreast of the major breakthroughs.

The earlier part of this chapter examined the signaling styles, I/O buffers, and basic topology designs. When all these are placed together, they constitute a system. As the I/O timing becomes more aggressive, second-order effects that were not so important now needed accounting. These will be discussed in the next section.

4.12.1.5. *SSO Reduction Using Low-Weight Coding*   The worst-case SSO impact occurs when all the bits toggle from one similar state to another state simultaneously. For example, eight bits all switching 1's to 0's will cause the worst-case ground bounce.

If one extra bit is added, interesting possibilities occur. The eight bits correspond to 256 patterns, whereas the nine bits correspond to 512 patterns. Using this ninth bit, instead of sending all 0's to an all-1 transition pattern, the outgoing data pattern is inverted (i.e., all 1's inverted to 0's; therefore there is no change in the outgoing driven data bits) but the ninth bit is sent as a 1 to the receiver as an indication that the inversion has occurred at the driver. The receiver detects an all-0 incoming pattern with an invert sign bit. It then inverts the received signal before passing it onto the logic. In this case, instead of eight bits toggling, only one bit toggled, reducing the SSO noise.

This can be extended to cases where not all bits toggle from one state to the other. If the number of bits changes state and if it is greater than 4, then the sign bit can be toggled. This action limits the maximum number of toggling bits to four, which in turn leads to 50% reduction in SSO noise [Nakamura, 1996, 1997].

This scheme can be understood with the help of Table 4-6. The current data driven from the core logic is compared to the previous state of the coded I/O

TABLE 4-6. Low Weight Encoding Scheme

Clock	Data from Core Logic	Difference	Invert	Ninth Bit	Coded I/O Drive Output
Initialize	0000 0000	0	No	0	0000 0000
1	1111 1111	8	Yes	1	0000 0000
2	0000 0111	3	No	0	0000 0111
3	0001 1000	5	Yes	1	1110 0111
4	0001 1000	8	Yes	1	1110 0111
5	0001 0111	4	No	0	0001 0111

Note: Scheme prevents all the bits in a bus transitioning from one state to the other simultaneously, thereby reducing the SSO noise induced.

drive. If the state differs with a magnitude greater than 4, then the invert bit is asserted and the data is inverted before being driven out.

4.13. SYSTEM MODELING

With the reduced timing budget at high-speed I/O, the following second-order effects need to be accounted for:

- Cross talk induced due to board traces running together for long lengths. Other major sources of cross talk are the connector where perfect shielding is hard to obtain and the package.
- Ground return loops can induce cross talk and discontinuities affecting the signal quality. In simple simulations these effects are not included but care must be taken during board layouts to ensure the shortest ground return paths between a signal and its ground (power) are available. This can be done by ensuring ground return vias close to the signal.
- Variations in the transmission line velocity ($\pm 10\%$) and impedance ($\pm 10\%$) will invariably play a part in the signal quality.
- Power supply delivery, both AC and DC, through connectors and packages will affect the signal.
- Reference voltage jitters required to compare the received input signals. Their jitter directly determines how much noise margin is available at the receiver as well as the timing window available.

Therefore, it is important to model the system in the high-speed I/O environment.

The system can be modeled in two ways: detailed transistor level leading up to the system and one using behavioral I/O buffers. A detailed transistor level model is helpful in designing individual nets and sizing the drivers, predrivers, and any compensation units. This is usually done using tools such as SPICE [Tuinenga, 1992]. However, for simulating many nets (a whole bus) this quickly becomes slow and very cumbersome. The simulation time required increases drastically. Further tools used to lay out and extract parameters from the mother board may not interface with silicon simulators easily. Therefore, a less refined silicon model can be used while increasing the complexity of the board level model, which may include the layout details of the bus. This can be achieved if the buffer is defined behaviorally. A behavioral model basically consists of current–voltage (I–V) data for a buffer as well as voltage–time (V–t) data. A simple behavioral model is discussed in Section 4.6, which has been used to illustrate the concept of stub length. The I–V information is provided by using a resistor , and the V–t information is provided using the ramp rate. This resistor and voltage ramp model is the simplest linear behavioral form, and for nonlinear devices, the I–V and V–t tables can be used. Another advantage of this behavioral model is that the silicon details are not visible or required for system modeling. Therefore, a system model can be provided to other parties without disclosing the exact process technology.

A particularly useful model is IBIS (I/O Buffer Information Specification). The model represents a specific I/O buffer. For each unique buffer, the I–V curve is extracted over all the possible pad voltages (including overshoots and undershoots). The I–V curve should provide a minimum, maximum, and typical number. In addition, to obtain the timing information, the buffer should be characterized using a fixed-output resistance and a V–t table obtained. In addition, the package pin definition should be provided in terms of R, L, and C. Therefore, each I/O pin can have a unique driver and package pin, or a unique driver combined with a generic package pin information, or a generic driver with unique package pin information. A sample IBIS file is shown in Figure 4-52.

Once the basic device has been defined, it can be incorporated into the system design. The system design will include the networks, the power supplies, and clock and reference voltages. The nets can be individualized and unique properties in terms of lengths, cross-coupling, and impedance can be assigned. Further, the net data can also be incorporated from PCB layout tools such that the exact board configuration is simulatable. The net extraction from the layout to a simulation is automated.

Some of the layout tools can simultaneously understand the PCB stack-up. The stack-up defines the ordering, number, distance of the power planes, and signal traces. Therefore, cross talk and impedance can also be automatically inserted into a simulation model.

A tool that is used routinely in experimental verification of I/O is the eye diagram (see Figure 4-53). An eye diagram is an overlay of many hundreds or thousands of sampled waveforms at a node superimposed on top of each other (voltage with respect to time). The eye diagram should capture the worst- and best-case patterns. The eye can be either measured on existing platforms or it can be simulated. During measurement only one node can be observed at a

[[IBIS VER]	2.1		
[File Name]	40505.ibs		
	Comment			
[Date]	7/23/97		
[File Rev]	1.1		

| 1.X Simulation data, 2.X correlated Si data, 3.X Mature data
|

[Copyright]	Copyright 1997,JKLM Corp.

|

[Component]	Trial_Component
[Manufacturer]	JKLM Corp.

|

[Package]

|

	Typ	Min	Max
R_pkg	100mOhm	50mOhm	150mOhm
L_pkg	4nH	3nH	5nH
C_pkg	4pF	3pF	5pF

| Package parasitics to be used when no specific parasitics are given for a pin
|

[Pin]	Signal name	Model name	R_pin	L_pin	C_pin
AD0	HAD0	PIN01			
AD1	HAD1	PIN02	25mOhm	2nH	4pF

| The specified R,L,C values supersede the default values for the AD1 pin
|

AD2	HAD2	PIN01
E0	Vcc	Power
E2	Vss	GND

[Model]	PIN01

| One model used in defining a buffer. Each buffer needs to have a model associated with it.
   The same model can be shared among a number of buffers such as AD0 and AD2.
|

Model_type	I/O

VINL = 0.8
VINH = 2.0
Vmeas = 1.5
Cref = 0.0

|

	Typ	Min	Max
C_Comp	3pF	2pF	4pF
[Voltage Range]	3.3V	3.0V	3.6V
[Temperature Range]	50	−10	110

|

Figure 4-52. Typical IBIS file showing the various sections required to complete a component definition.

[Pulldown]
| *describes the pull down I–V, from −Vcc to twice Vcc. The negative pad voltages cause*
| *the substrate-pad diode to forward bias whereas the positive voltages reflect the pull*
| *down action.*
|

Voltage	I(typ)	I(min)	I(max)
−3.3	−300mA	−280mA	−320mA
−3.1	−260mA	−240mA	−280mA
...			
...			
6.6	50mA	48mA	52mA

|
[Gnd Clamp]

−3.3	1A	NA	NA
−3.1	0.93A	NA	NA
...			
...			
3.3	1.3pA	NA	NA

|
[Ramp]
|*Provides the ramp rate ofthe pull up and pull down operation.*
|
R_Load = 65 Ohm

	Typ	Min	Max
dv/dt_r	1.1V/300ps	NA	NA
dv/dt_f	1.2V/350ps	NA	NA

|
[Falling Waveform]
|*Provides the volage-time table for the pull down action.*
|
V_fixture = 3.3V
V_fixture_min = 3.0V
V_fixture_max = 3.6V
R_fixture = 50.0Ohm

Time	V(Typ)	V(min)	V(max)
0.0	3.3	NA	NA
20E-12	3.5	NA	NA
40E-12	3.05	NA	NA
60E-12	2.85	NA	NA
...			

Figure 4-52 (*Continued*)

time, but many patterns can be run. In each cases the random data patterns are the most beneficial. Similarly the eye can be measured for different power supply voltages, temperatures, and process conditions and then a cumulative eye assembled.

The eye can also be simulated. This can be done by first creating a complete network, including silicon and the PCB. Then various parameters can be varied

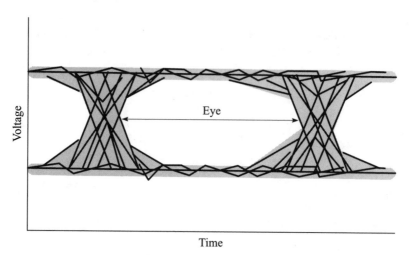

Figure 4-53. Eye diagram that captures many waveforms at a node that are then superimposed. The gray area is the uncertain area where there is no guarantee that a signal can be reliably captured. The white area represents the area where the signal is stable and can be captured. The eye diagram can be measured or simulated.

and all the results superimposed on a plot to create an eye diagram. Some parameters that can be varied are the data patterns (with and without SSO and random), PLL and clock jitters, voltage supplies, cross talk, ISI, termination and driver strengths, PCB line and stub lengths, impedances and velocity propagation, package line and stub lengths, impedances and velocity of propagation, and silicon parameters for both the driver and receiver. After creating a large matrix of all the relevant parameters, the simulation is carried out and the results are superimposed. The eye can then be measured and the results are carried over to the noise margin and timing calculations.

4.14. I/O INFORMATION ON THE INTERNET

A huge amount of very specific information exists on manufactured IC I/O. These include drivers, receivers, and system considerations and are usually good sources to obtain practical data. In addition, other I/O and ESD-related information such as ESD products, connectors, sockets, packaging, PCB manufacturing, and test equipment can be obtained from the Internet.

A typical method is to "search" the Internet for the keyword(s). These will usually yield a long list of results. A good idea is to read the list in the reverse chronological order.

Alternatively, if the relevant manufacturer is known, then their web site can be reached (e.g., http\\www.intel.com or http\\www.hitachi.com) and the I/O information can be obtained from the product information and specification section. These are generally listed under the silicon products, IC heading. If the manufacturer address is not known, it can be first "searched" and then contacted.

4.15. SUMMARY

The following are some of the major considerations discussed in this chapter:

- The most convenient form of communication between chips is the PCB transmission line, and on-chip, the metal oxide *RC* line. The transmission velocity of a transmission line is dependent only on its dielectric constant.

- Transmission line phenomena are observed when the rise time is smaller than 2.5 times the TOF in a line.

- A lumped capacitor model can be used when the rise times are larger than five times the TOF.

- Two types of termination strategies are typically used: parallel (GTL) and serial (CMOS). A parallel-terminated net will usually have better settling time than a series-terminated line.

- Two types of signal transmission strategies are typically used: common clock and cotransmitted clock. The CTC avoids the penalty of TOF.

- Topology affects the speed considerably. The fastest topologies are generally point to point and multiagent nets are the slowest. The multiagent buses are slow primarily because of the longer stub lengths required to connect the multiple agents, forcing the edge rates to be slow. Also the multiple agents force the bus to be long and nonsymmetric, thus making settling times long.

- The large process, temperature, and voltage variations cause significant variations in I/O buffer characteristics. Compensation is therefore essential for fast buses. The output resistance and the slew rates can be compensated.

- There are primarily two methods employed to compensate I/O buffers: analog and digital techniques.

- A major concern in I/O design is the SSO noise due to the large number of I/O buffers switching in one direction. Methods to control this noise involve slowing down the predriver, compensating the output driver strength, and increasing the number of power supply wires.

- A crucial tool in I/O design is the vast amount of information available on the Internet. Before starting a new system design, it is well worth the time to quickly browse through the options available.

REFERENCES

[Allen, 1987] P. E. Allen, D. R. Holberg, *CMOS Analog Circuit Design*, Holt, Reinhart and Winston, Orlando, FL, 1987, p. 231.

[Alvarez, 1995] J. Alvarez, H. Sanchez, G. Gerosa, and R. Countryman, "A Wide Band Low-Voltage PLL for PowerPC^{TM} Microprocessors," *IEEE JSSC*, **E78-c** (6), 1995, p. 631.

[Annaratone, 1986] M. Annaratone, *Digital CMOS Circuit Design*, Kluwer, Boston, 1986.

[Bakogolu, 1990] H. B. Bakoglu, *Circuits, Interconnections and Packaging for VLSI*, Addison-Wesley, 1990.

[Catlett, 1992] C. E. Catlett, 'Balancing Resources," *IEEE Spect.*, Sept 1992, p. 48.

[Chandrakasan, 1994] A. Chandrakasan, A. Burstein, and R. W. Brodersen, "A Low Power Chipset for Portable Multimedia Applications," *Proc. ISSCC*, 1994, p. 82.

[Clark, 1989] R. M. Clark, *Printed Circuit Engineering*, Van Nostrand Reinhold, New York, 1989.

[Dally, 1996] W. Dally and J. Poulton, "High Performance Signaling, How to Transmit a Gb/s per Pin," Hot Interconnects Symp., Tutorial, 1996.

[Declercq, 1993] M. J. Declercq, M. Schubert, and F. Clement, "5V-to-75V CMOS Output Interface Circuits," *ISSCC*, 1993, p. 162.

[DeHon, 1993] A. DeHon, T. Knight, and T. Simon, "Automatic Impedance Control," Digest of Technical Papers, *ISSCC*, 1993, p. 164.

[Deutsch, 1994] A. Deutsch, G. Arjavalingam, C. W. Surovic, A. P. Lanzetta, K. E. Fogel, F. Doany, and M. B. Ritter, "Performance Limits of Electrical Cables for Intrasystem Communication," *IBM J. Res. Dev.*, **38** (6), 1994, p. 659.

[Flatt, 1992] M. Flatt, *Printed Circuit Board Basics*, Miller Freeman, San Francisco, 1992.

[Gabara, 1992] T. J. Gabara and S. C. Knauer, "Digitally Adjustable Resistors in CMOS for High-Performance Applications," *IEEE JSSC*, **27** (8), 1992, p. 1176.

[Gabara, 1997] T. J. Gabara, W. C. Fischer, J. Harrington, and W. W. Troutman, "Forming Damped LRC Parasitic Circuits in Simultaneously Switched CMOS Output Buffers," *IEEE JSSC*, **32** (3), 1997, p. 407.

[Gunning, 1992] B. Gunning, L. Yuan, T. Ngyen, and T. Wong, "A CMOS Low-Voltage-Swing Transmission Line Transceiver," *Proc. ISSCC*, 1992, p. 58.

[Guo, 1994] B. Guo, A. Hsu, Y. C. Wang, and J. Kubinec, "A 125 Mbs CMOS All-Digital Data Transceiver Using Synchronous Uniform Sampling," *Proc. ISSCC*, 1994, p. 113.

[Hanafi, 1992] H. I. Hanafi, R. H. Dennard, C.-L. Chen, R. J. Weiss, and D. S. Zicherman, "Designing and Characterization of a CMOS Off-chip Driver/Receiver with Reduced Power-Supply Disturbance," *Proc IEEE JSSC*, **27** (5), 1992, p. 783.

[Hu, 1994] C. Hu, "Low-Voltage CMOS Device Scaling," *Proc. ISSCC*, 1994, p. 87.

[Intel, 1996] Pentium Pro Family Developers Manual, No. 242690, Intel Corp, 1996, p. 12-3.

[Ishihara, 1994] N. Ishihara and Y. Akazawa, "A Monolithic 156 Mb/s Clock and Data-Recovery PLL Circuit Using the Sample-and-Hold Technique," *Proc. ISSCC*, 1994, p. 111.

[Lee, 1994] T. H. Lee, K. Donnelly, J. Ho, J. Zerbe, M. Johnson, and T. Ishikawa, "A 2.5V Delay Locked Loop for an 18Mb 500MB/s DRAM," *IEEE ISSCC*, 1994, p. 236.

[Long, 1990] S. L. Long and S. E. Butler, *Gallium Arsenide Digital Integrated Circuit Design*, McGraw-Hill, New York, 1990, Chapter 5, p. 245.

[Millman, 1987] J. Millman and A. Gabriel, *Microelectronics*, McGraw-Hill, New York, 1987.

[Mooney, 1995] R. Mooney, C. Dike, and S. Borkar, "A 900 Mb/s Bidirectional Signaling Scheme," *IEEE JSSC*, **30**(12), 1995, p. 1538.

[Nakamura, 1996] K. Nakamura and M. A. Horowitz, "A 50% Noise Reduction Interface Using Low-Weight Coding," *Symp. VLSI Circuits, Digest of Tech. Papers*, 1996, p. 144.

[Nakamura, 1997] K. Nakamura, K. Takeda, H. Toyoshima, K. Noda, H. Ohkubo, T. Uchida, T. Shimizu, T. Itani, K. Tokashiki, and K. Kishimoto, "A 500 MHz 4Mb CMOS Pipeline-Burst CACHE SRAM with Point-to-Point Noise Reduction Coding I/O," *IEEE Intl. Solid-State Circuits Conf.*, 1997, p. 406.

[Nogami, 1994] K. Nogami and A. E. Gamal, "A CMOS 160 Mb/s Phase Modulation I/O Interface Circuit," *Proc. ISSCC*, 1994, p. 109.

[Ooishi, 1995] T. Ooishi, Y. Komiga, K. Hamade, M. Asahura, K. Yasuda, K. Furutami, H. Hidaka, H. Miyamoto, and H. Ozaki, "An Automatic Temperature Compensation of Internal Sense Ground for Subquarter Micron DRAMs," *IEEE JSSC*, **30**(4), 1995, p. 471.

[Pilo, 1996] H. Pilo, S. Lamphier, F. Towler, and R. Hee, "A 300 MHz 3.3 V 1 Mb SRAM," *Proc. ISSCC*, 1996, p. 148.

[Prince, 1994] B. Prince, "Memory in the Fast Lane," *IEEE Spect.*, Feb., 1994, p. 38.

[Rainal, 1994] A. J. Rainal, "Eliminating Inductive Noise of External Chip Interconnections," *IEEE JSSC*, **29**(2), 1994.

[Rambus] RDRAM®, Rambus Inc., Mountain View, CA.

[Reese, 1994] E. A. Reese, H. Wilson, D. Nedwek, J. Jex, M. Khaira, T. Burton, P. Nag, H. Kumar, C. Dike, D. Finan, and M. Haycock, "A Phase-Tolerant 3.8 GB/s Data-Communication Router for a Multiprocessor Supercomputer Backplane," *Proc. ISSCC*, 1994, p. 296.

[Rein, 1996] H. M. Rein and M. Moller, "Design Consideration for Very-High-Speed Si-Bipolar IC's Operating up to 50GBs," *IEEE JSSC*, **31**(8), 1996, p. 1076.

[Rizzi, 1988] P. A. Rizzi, *Microwave Engineering Passive Circuits*, Prentice-Hall, Englewood Cliffs, NJ, 1988, p. 2.

[Satyanarayana, 1995] S. Satyanarayana and K. Suyama, "Resistive Interpolation Biasing: A Technique for Compensating Linear Variation in an Array of MS Current Sources," *IEEE JSSC*, **30**(5), 1995, p. 595, and references therein.

[Sekiguchi, 1995] T. Sekiguchi, M. Horiguichi, T. Sakata, Y. Nakagome, S. Ueda, and M Aoki, "Low Noise, High Speed Data Transmission Using a Ringing-Canceling Output Buffer," *IEEE JSSC*, **30**(12), 1995, p. 1569.

[Senthinathan, 1994] R. Senthinathan and J. L. Prince, *Simultaneous Switching Noise of CMOS Devices and Systems*, Kluwer Academic, Boston, 1994.

[Shirotori, 1991] T. Shirotori and K. Nogamir, "PLL Based Impedance Control Output Buffer," *Symp. on VLSI Circuits*, 1991, p. 49.

[Stone, 1983] H. S. Stone, *Microcomputer Interfacing*, Addison-Wesley, Reading, MA, 1983, Chapter 5, p. 161.

[Takahashi, 1995] T. Takahashi, M. Uchida, T. Takahashi, R. Yoshimo, M. Yamamoto, and N. Kitamura, "A CMOS Gate Array with 600 Mb/s Simultaneous Bidirectional I/O Circuits," *IEEE JSSC*, **30**(12), 1995, p. 1544.

[Tuinenga, 1992] P. W. Tuinenga, *A Guide to Circuit Simulation & Analysis Using Pspice*, Prentice-Hall, Englewood Cliffs, NJ, 1992.

[Weste, 1985] N. H. E. Weste and K. Eshraghian, *Principles of CMOS VLSI Design*, Addison-Wesley, Reading, MA, 1985.

[Yamauchi, 1996] T. Yamauchi, Y. Morooka, and H. Ozaki, "A Low Power High Speed Data Transfer Scheme with Asynchronous Compressed Pulse Width Modulation for AS-Memory," *IEEE JSSC*, **31**(4), 1996, p. 523.

[Yang, 1996] Y. Yang and J. R. Brews, "Design for Velocity Saturated Short-Channel CMOS Drivers with Simultaneous Switching Noise and Switching Time," *IEEE JSSC*, **31**(9), 1996, p. 1357.

CHAPTER 5

LAYOUT ISSUES

It has been observed repeatedly that an incorrect layout can jeopardize ESD protection [Maloney, 1992, 1993; Wei, 1993]. The ESD current and voltage stresses are very efficient in finding the weakest spots in a circuit and destroying them. This makes it essential to have a very robust layout of all the core and I/O circuits. However, in this book we have stressed the creation of ESD paths (designed) with low impedance that channel current away from other high-impedance (random) paths that would otherwise break down and be damaged. This diverting action significantly eases layout guidelines for the core circuits and allows designers to concentrate on the designed ESD paths.

Creation of an ESD path also assumes that each I/O component is basically robust. The ESD zap path only guarantees a safe channel to conduct the stress and assumes that the basic devices meet a certain minimum stress level. One of the most sensitive elements susceptible to ESD stress is the NMOS output transistor [Carbajal, 1992]. The NMOS output device layout and associated options are therefore discussed first. Other concerns are metal wiring widths, placement of antenna diodes, ESD diodes, and power clamps. These issues and the related layout methods are discussed in this chapter.

5.1. OUTPUT TRANSISTOR LAYOUT

The NMOS transistor is most sensitive to ESD because of the NPN snapback action that gets initiated at low voltages and, if not properly designed, then rapidly enters into second breakdown. A layout design such as the one in Figure 5-1 works well for nonsilicided technologies. In this structure, the dis-

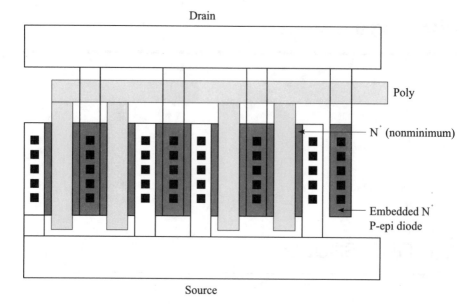

Drain

Poly

N' (nonminimum)

Embedded N'
P-epi diode

Source

Figure 5-1. Nonsilicided NMOS "ladder" output transistor layout in unsilicided technology.

tance of the drain diffusion to the gate and the number and placement of contacts are important [Stricker, 1995; Amerasekera, 1995, p. 59]. The N^+ diffusion with its high resistivity acts as a ballast resistor, preventing any one finger from hogging all the current and consequently preventing other fingers from breaking down. To best use the N^+ ballast resistivity, the placement of contacts is important. Usually the contacts are placed at a nonminimum distance from the N^+/polysilicon edge to create the resistance.

The "waffle" NMOS or PMOS layout is another option and is shown in Figure 5-2. The waffle layout option can be fairly compact: therefore it is more area efficient than a ladder layout and has comparable ESD performance [Baker 1989a; Maene, 1992]. Another advantage is that the drain capacitance can be lowered by a factor of 4 because each minimum drain area is bounded by a polysilicon gate on four sides. This can be seen when Figures 5-1 and 5-2 are compared [Razavi, 1995]. Despite its area efficiency, it has certain device modeling issues. The width-to-length ratio is not well defined near the corners in this waffle, and each cell must be modeled separately and treated as a unit.

For example, this has been done by Baker [1989a] for a waffle structure with a right-triangle corner. Since the triangular corner has variable gate length, an effective value has to be calculated:

$$\Delta W_{\text{eff}} = \frac{4L}{\pi} \ln \frac{L + W_s}{L} \qquad (5\text{-}1)$$

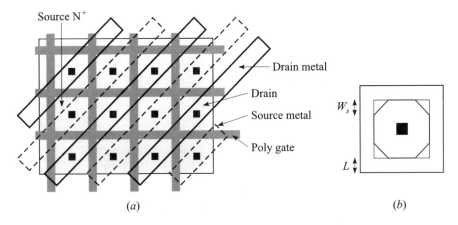

Figure 5-2. (*a*) Waffle NMOS transistor structure and (*b*) one individual cell [Baker, 1989a].

where W_s is the length of one side of the right triangle in the corner and L is the gate length.

Since the polysilicon is needed to isolate the source and drain, each gate ring must be complete. The physical implementation and modeling together impose some limitation in granularity (usually small). Also the current conduction is not uniform in the device during both ESD and normal operating regimes.

It is crucial to have a well-balanced and uniform layout. Effects of nonuniform contact and via spacing can be detrimental due to current crowding. Krakauer [1992] describes a layout error, shown in Figure 5-3, that reduces a thick-field grounded-gate device with robustness from 7000 to 3500 V. This deterioration occurred due to via 2 placement (by error) in only one area; thereby only two fingers of the TFGG were utilized, as opposed to five which were laid out. This highlights that it is good practice to keep the layout as uniform as possible, which encourages equal conduction of the ESD current.

Consider another layout of the NMOS as shown in Figure 5-4, where the current is sourced by the metallization from one side only, the current is distributed poorly, and damage to the crowded side is possible. A better layout practice is shown in option (*b*), where the current distributes more evenly and higher device ESD performance should be expected [Wada, 1995].

In keeping with the ESD path method to provide a robust current path from V_{scc} to the pad, additional embedded diodes can be added, as shown in Figure 5-1. The embedding allows uniform voltage on all transistor drains. Typically these diodes are added to reinforce the existing N^+-(drain)-to-P-epi diodes. If sufficient drain edge already exists, as would happen for large drivers, no additional diodes are needed.

A similar layout can be used for the PMOS transistor, as shown in Figure 5-5. Here the PNP snapback effect is not pronounced or destructive. Ballasting

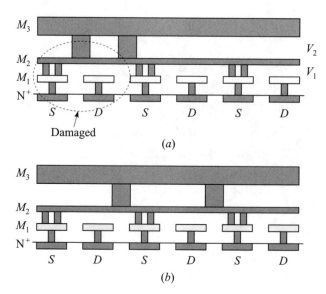

Figure 5-3. Early damage of robust Si structure due to via nonuniformity: (a) damaged transistor and (b) correct layout. This damage occurred because of nonuniform via 2 placement [Krakauer, 1992].

is usually not required, and minimum geometry device is allowable (i.e., with proper hot-electron considerations that are not as significant as for the NMOS device). When the diodes are integrated into the PMOS device, the pad nodes should face the N^+ of the diode. This reduces the diode resistance. If the drain is placed in the center, as shown in Figure 5-5c, with the source interposed between the pad and the N^+ of the diode, the current has to traverse a longer path, thereby increasing the diode resistance. Also, the P^+ source creates a

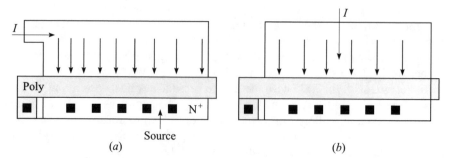

Figure 5-4. (a) Weak spot in grounded-gate structure due to nonuniform current distribution and (b) a more uniform current distribution by centering the source of the current [Hulett, 1981].

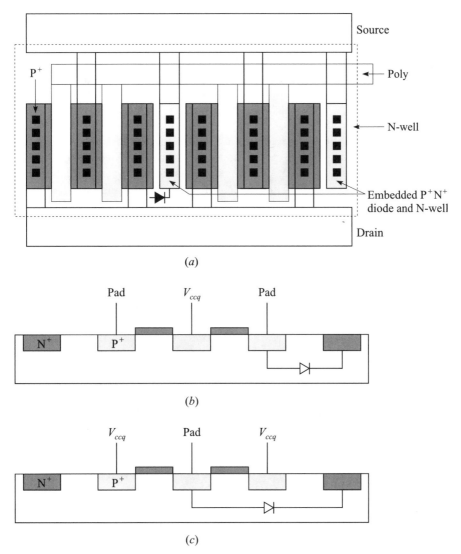

Figure 5-5. (*a*) A PMOS transistor layout showing the embedded diode and an N-well pickup. (*b*) Cross-sectional view showing relative pad (drain), V_{ccq} (source) and N^+ of diode. (*c*) Placing the pad in the center creates a high-resistance path for the ESD diode; therefore, this layout should be avoided. In this layout P^+ also creates a depletion layer in the N-well, thus reducing the available area for the current flow during the diode conduction.

depletion layer reducing the available N-well area for the diode current. This further increases the diode resistance and will prevent effective diode clamping.

Typically, the N-well contacts are sufficient to allow good transistor biasing but are usually not well strapped to conduct heavy ESD currents. This is in contrast to the NMOS P-epi body where the body is in solid contact with the P^+ substrate which is usually in contact with a good ground connection on the package. To construct an effective current path from a pad to V_{ccp}, a deliberate diode has to be inserted between the two. These are shown as the embedded diode. The diode can be distributed evenly, and usually the PMOS size is large enough so that the diode is not a prohibitively large percentage area of the device.

The NMOS layout styles worked well before the introduction of silicides. With silicidation the NMOS performances plummeted [Amerasekera, 1995, p. 65]. The ballasting action of N^+ diffusion was lost with corresponding loss of ESD performance [Polgreen, 1992]. The ballast resistance must now be incorporated explicitly or other techniques such as GCNMOS (gate-coupled NMOS) or RGNMOS (ratioed-gate NMOS) must be employed to help evenly break down the NMOS fingers. The GCNMOS and RGNMOS have been discussed earlier so now the ballast option is considered.

The question arises of where to place the ballast resistor. Should it be at the source or at the drain. Experience has shown that the resistors on the drain are more effective [Polgreen, 1992]. Some reasons are as follows (refer to Figure 5-6):

- When the resistor is placed between the drain and pad, the drain voltage is reduced (by IR drop) during an ESD zap, thereby reducing the gate stress [Ramaswamy, 1995]. When the resistor is in between the source and the V_{ssp} power rail, the drain voltage is at the pad level, and gate-to-drain overvoltage damage can occur.

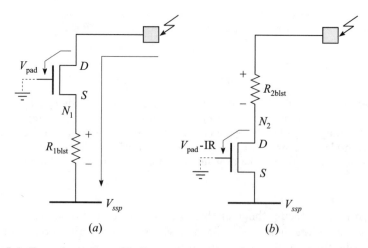

(a)   *(b)*

Figure 5-6. Two placements of ballast resistors (*a*) at the source and (*b*) at the drain of the NMOS device.

- During snapback and second breakdown, a significant portion of the current path (up to 25%) is directly from the drain to the substrate. By placing the resistor on the drain side, the drain substrate current is forced through the resistor, leading to better resistor utilization and higher *IR* drop.
- As discussed earlier, the forward biasing of the substrate to N^+ source lowers the V_{t1} below V_{t2} levels, enabling all the fingers to trigger on and share the current (see Section 3.2.1). By placing the resistor in the source side, this forward biasing is lowered; therefore, the V_{t1} modulation effect is weakened.

The implemented layout is shown in Figure 5-7, where explicit N-well resistors have been added [Carbajal, 1992; Tong, 1996]. The benefit of the N-well can be judged from the data provided by Tong [1996], which shows that an NMOS (500/1.0) could sustain only a 1.4-kV HBM zap, whereas an NMOS (500/0.6; $50/0.6 \times 10\,\text{legs} = 500/0.6$) with 50Ω-per-leg N-well ballast resistor could enhance the HBM performance to 2.9 kV. The N-well should be separate so

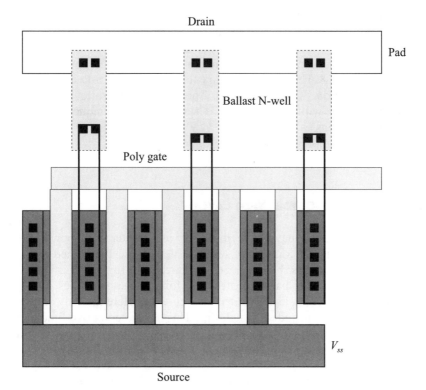

Figure 5-7. Silicided NMOS layout showing use of external ballast.

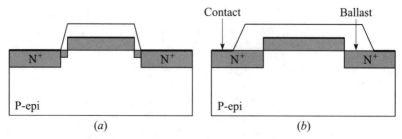

Figure 5-8. (*a*) Current silicidation process used in core logic allows no ballasting action of diffusion; (*b*) by selective blocking of the silicide (in I/O) a ballasting resistor can be created.

that all the drains are different nodes. Usually N-well additions are expensive in area as the N-well has large spacing requirements.

If the N-well is used as an external ballast, it complicates the layout sufficiently that a waffle NMOS layout will be difficult [Baker, 1989a]. However, the PMOS device, which does not need an N-well ballast, still retains that option. Since the PMOS is usually larger (having lower mobility of holes), the reduction of drain capacitance by waffling may be significant.

Another solution to this ballast issue is to block the silicidation process for ESD-affected devices in the relevant diffusion areas. In this scenario the core circuits can benefit from regular silicidation and the NMOS output driver circuits can experience a robustness similar to a nonsilicided process. This is shown in Figure 5-8. Here, it is necessary to have an extra S-D implant before the spacer definition [Krakauer, 1992]. The ballast action allows device ESD robustness scaling with device width. Using such a technique, a process is developed which passes HBM ESD of 3 kV in the first silicon. The only drawback is the extra implant and masking operation required.

5.2. THICK-FIELD-OXIDE (TFO) LAYOUT

Grounded-gate devices have been discussed in Chapter 1 as power clamps utilized in previous generations of technologies. These functioned well with nonsilicided technologies, but their performance has dropped drastically with newer technologies. The basic construction follows the guidelines of NMOS output driver transistors, as illustrated in Figure 5-9. Here the only addition is the small thin-oxide grounded-gate device. This device serves two functions: to act as the local clamp to an input buffer and as a triggering device to the thick-oxide grounded-gate device. The only requirement is that the two devices be located physically close to each other so they can interact. In the layout in Figure 5-9 both devices share a common source to allow area reduction as well as good coupling. The thin-oxide grounded gate needs a resistor to limit the

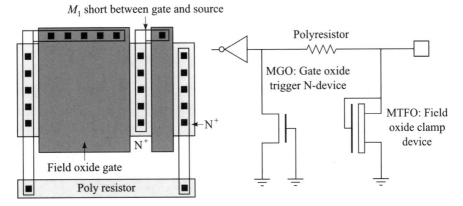

Figure 5-9. Layout of TFO device showing thin-oxide grounded-gate device to trigger TFO.

current flowing into it during an ESD event and to protect it from damage [Maloney, 1988; Duvvury, 1991].

5.3. DIODE LAYOUT

Diodes are an essential part of ESD whether they are deliberately placed circuit elements or parasitics. In this version of the current path methodology diodes have been extensively used as they are simple to design. They are also very forgiving. In Figure 5-10 a diode laid out in the N-well is shown. These diodes are used when isolation is required from the substrate, as in V_{ccp}-to-V_{cc} clamps. The N^+ and P^+ areas should be minimized and their peripheries maximized. The areas contribute capacitance and the edges contribute conductance. The diodes in CMOS technologies are really PNP transistors, which means that the current into the emitter is larger than the one in the base. Thus the P^+ emitter

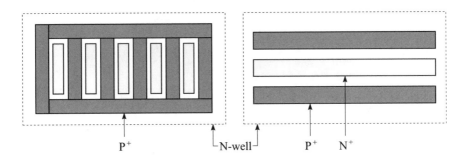

Figure 5-10. Two styles of diode layouts. Note in both the emitter area is larger than the base.

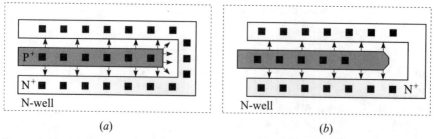

Figure 5-11. (*a*) Diode corners are susceptible to ESD damage as they conduct more current due to three-dimensional implant profiles and (*b*) a simple fix to the high current by increasing the resistance at the corner.

periphery should be made larger than the N^+ base. The N^+-to-P^+ spacing should be kept minimum to reduce the resistance of the diode. Similarly robust and equal metallization, contacts, and vias should be placed to equalize the current [Voldman, 1992; Jaffe, 1990]. The diode P^+ stripe should be greater than $100\,\mu m$ to ensure both ESD and EOS reliability [Voldman, 1994a; Ramaswamy, 1996].

If diodes are scaled down very aggressively, then second-order effects can become important in determining their robustness [Voldman, 1992; Ramaswamy, 1996]. Consider the effect of a three-dimensional implant profile and the lowered diode resistance due to current spreading at the corners on the diode current. The corner current may be $\sim 50\%$ higher than the current along the diode edge [Voldman, 1992]. The diode will burn out at these corners unless some other soft spot breaks down sooner. To compensate for this enhanced conduction, it is necessary to increase the diode resistance near the corner, thereby reducing the current, as shown in Figure 5-11. This can be done by

- removing the P^+ contacts near the P^+ corner,
- removing the N^+ contacts near the P^+ contacts,
- beveling the corner, and
- increasing the space between P^+ and N^+ at the corner.

5.4. DECOUPLING CAPACITORS

Decoupling capacitors are increasingly becoming more important to reduce the voltage ripple on the power supply of high-performance chips. Trends indicate that two approaches are being evaluated. One is to provide more decoupling (more area and higher dielectric constant) and the other is to reduce the power requirements (low power circuits and lower dielectric constant for the interconnect). This trend is shown in Figure 5-12.

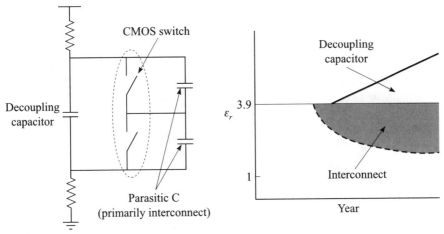

Figure 5-12. Coming split in on-chip interconnect and decoupling capacitor in dielectric materials [Miyamoto, 1997; Jeng, 1973; Paraszczak, 1993; Lee, 1997]: (*a*) Charge sharing between local decoupling capacitor and interconnect capacitance. (*b*) Method to minimize effect of charge sharing by making decoupling capacitor much larger than interconnect capacitance. The decoupling capacitor can be made larger using larger area, reduced gate oxide thickness, or increased dielectric constant.

The capacitance required for a chip can be estimated using Eq. 5-2. For example, for 2.5-V technology with a voltage droop of no more than 5% (0.125 V) and a current of 10 A switched in 4 ns (250 MHz) a capacitor greater than 3.2 nF is needed. In reality, a peak current of 40 A may be drawn in under 500 ps, which makes the capacitor required to be 100 nF. As the voltage decreases, the chip size increases, and the frequency increases, this capacitance value will increase further. A layout of such a simple decoupling capacitor based on an NMOS transistor is shown in Figure 5-13. The two terminals are the top polysilicon connection (to V_{cc}) and the inversion layer connected to V_{ss}:

$$Q = I\Delta t = C\Delta V \qquad C \geq \frac{I\Delta t}{\Delta V} \qquad (5\text{-}2)$$

Since the capacitance scales directly with area, efficient layout is crucial. This would also mean maximization of the gate area and reduction of the overheads such as the N^+ diffusion needed to make contacts. However, there are constraints. Borrowing from simple MOS theory, the RC time constant can be written as

$$RC = \left(\frac{1}{q\mu EWt}\right)(C_0 WL) = \left(\frac{L}{q\mu V_{res}}\right)(C_0 L); \qquad RC\alpha L^2 \qquad (5\text{-}3)$$

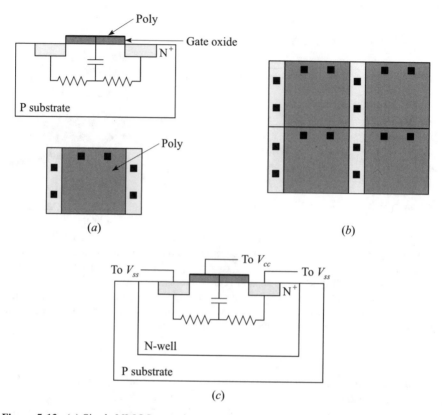

Figure 5-13. (a) Single NMOS capacitor cell and (b) arrayed capacitor cell. Capacitance can also be made in N-well. (c) Electrical model.

where L is half the capacitor length, t is the inversion layer thickness, W is the capacitor width, and V_{res} is the maximum allowable drop on the resistor. The RC time constant should be less than the greatest frequency of interest. This also indicates that the frequency roll-off is proportional to the square of the length.

With increasing clock frequencies in VLSI, the transients are getting sharper (i.e., rise and fall times being about one-tenth of the period). To respond to these fast demands of charge, the cutoff frequency of the capacitor must be large, which means that it cannot be very long. Typically the length should be kept to a few micrometers. For larger lengths, the electrons are not able to respond deep in the center of the capacitor, and they are not effective for decoupling the high frequency.

If a simple calculation is made in terms of electron velocity, the saturation velocity, and the saturation electric field, then the half length of a capacitor able to respond is given as

$$\frac{L}{2} \le \frac{(V_{dc})/10}{E_{sat}} = V_{sat}\frac{1}{10f} \tag{5-4}$$

where $V_{sat} = 1 \times 10^7$ cm/s, $E_{sat} = 1 \times 10^4$ V/cm ($\mu = 1000$ cm^2/Vs), and $V_{dc}/10$ is 10% of the technology voltage allowed to drop in the capacitor. In addition, $1/10f$ represents the rise and fall times of the signal as a function of the clock period and the time allowed for charge to come from the channel to the power distribution metal. To get a feeling for this length, consider a 2.5-V technology operating at 250 MHz; then the maximum gate length is 8μm. Beyond 8μm an electron will not be able to respond in time to respond to the edge.

Another way to look at the decoupling capacitor is from a distributed RC point of view. Here the input impedance of the capacitor is seen as [Larson, 1997]

$$Z_{in} = \sqrt{\frac{(R/4)}{sC}} \coth\left(\sqrt{\frac{sRC}{4}}\right) \tag{5-5}$$

which can be expanded into a series as in

$$Z_{in} = \frac{1}{sC} + \frac{R}{12} - \frac{sR^2C}{720} \tag{5-6}$$

For a lumped RC model the input impedance is given as

$$Z_{in\ lumped} = R + \frac{1}{sC} \tag{5-7}$$

Comparing the lumped and the distributed models, the equivalent resistance is $R/12$ of the channel resistance between the source and the drain of an equally sized MOS device. The equivalent resistance also shows a frequency-dependent behavior (see Figure 5-14).

The PMOS can act as a good low-resistance capacitor. The N-well resistance is considerably smaller than the P-epi resistance. Thus better performance can be expected. However, in this case the N-well is tied to V_{ss}. Generally, the N-well is tied to the highest potential in a CMOS circuit. The only caution here is to ensure that the V_{ss} N-well is not close to another N-well tied to the V_{cc} to prevent NPN snapback breakdown.

One area that can be used to yield high-quality capacitors is the space under the wire bond pad. Here, the pad metal to metal or any another metal (not the pad) attached to polysilicon capacitors can be formed [Ker, 1996]. Although each capacitor may not be large [i.e., rough calculation shows that a 500-nm interlayer dielectric (ILD) thickness has a capacitance 50 times less than an equally sized 10-nm oxide capacitance], the number of pads on a large chip makes it a sizable quantity. These options are shown in Figure 5-15.

Two key requirements for capacitor layouts are:

- high area efficiency and
- easy replication and arraying.

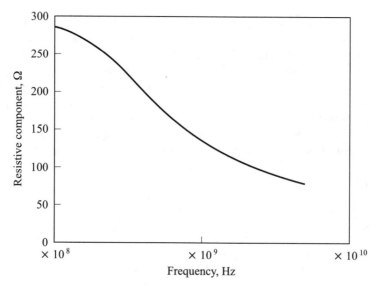

Figure 5-14. Equivalent resistance of decoupling MOS capacitor serves as a function of frequency. At higher frequency the capacitor impedance decreases, allowing more current to pass closer to the source and drain taps, thereby reducing resistance.

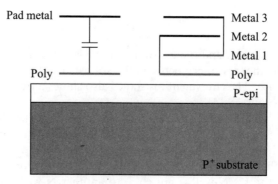

Figure 5-15. Capacitors that can be formed under a pad are (*a*) pad metal and poly and (*b*) an interleaved metal layer sandwich for efficient layout. The stress generated during wire bonding imposes constraints in the capacitance construction [Ker, 1996].

5.5. SCR LAYOUT

The SCR functionality is extremely sensitive to variations in the fabrication process. For each process the correct layout is found by experimenting on a large matrix of SCR devices [Carbajal, 1992; Ker, 1992; Amerasekera, 1995, p. 116; Diaz, 1994; Duvvury, 1995]. The one that meets all the trigger voltage, leakage, and I–V requirements is then selected. For these reasons, they are not easily scaleable or mapped to newer technologies directly. In Figure 5-16, three SCR layouts are shown corresponding to an SCR, an MVTSCR, and an LVTSCR. For working details the reader is referred back to Section 2.2.3–2.2.5.

5.6. ANTENNA DIODE LAYOUT

In section 3.6, issues related to antenna diodes were briefly explained; here gate capacitance and N^+/P diode options are discussed as solutions to the problem. The layouts of some diffusion types of antenna diodes are shown in Figures 5-17 and 5-18. In these figures three options are shown that need attention. Any time there is a power-supply-to-power-supply N^+-P-N^+ possibility, there is a chance for significant current to flow and cause damage. The electrical equivalent circuits have already been shown in Figure 3-16 in Section 3.6. Some protection can be obtained by adding an N-well around an antenna diode (expensive in area) by increasing the resistance in the path. However, this is only a partial solution. The satisfactory method should be to prevent the core voltages from reaching the NPN snapback voltage.

5.7. RESISTOR LAYOUTS

There are three types of resistors available in modern CMOS technology: N-well, pinched N-well, and polysilicon. These options are shown in Figure 5-19. In older technologies with no silicide, the N^+ or P^+ could be used as a resistor as well. However, even in these unsilicided technologies the diffusion resistors suffered from high capacitance and lower breakdown (reverse bias to substrate) handicaps.

The construction and layout of these devices are shown in Figure 5-19. Consider the N-well resistor. Typical values of sheet resistivity are 200–500 $\Omega/\square$ [Voldman, 1994a,b]. Thus resistors up to a few kilohms can be easily constructed using the N-well. Since the N-well has the coarsest features on the die, the area penalty can be prohibitive for large-valued resistors. If one end of the N-well resistor is tied to a pad, there is a chance of forward biasing the N-well to-P-epi diode, which may be a latchup hazard. For this reason, it is good practice to guard ring the resistor. The maximum voltage capability of the resistor is determined by the N-well-to-P-epi breakdown voltage.

Higher values of the resistor can be formed by pinching off the N-well [Ghandhi, 1977; Millman, 1987, p. 197]. This is done by placing a P^+ over the N-well. The depletion layer effectively reduces the cross-sectional area of the resistor, thereby increasing the resistance. By the same token, the N-well area is modulated as the high-voltage end very much like JFET devices. Consequently, this pinched N-well resistor is not a constant-resistance element and should be used with caution. For example, it has been utilized as an ESD protection

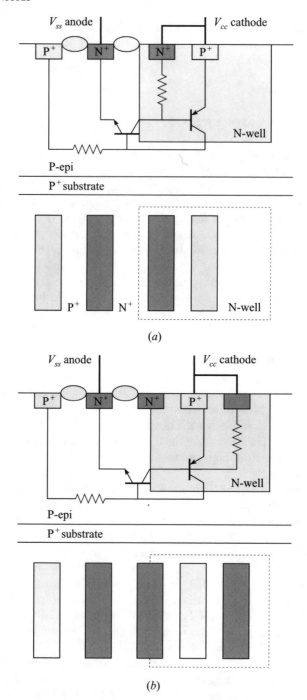

Figure 5-16. The SCR layout showing the forms of (*a*) a typical SCR, (*b*) an MVTSCR, and (*c*) an LVTSCR.

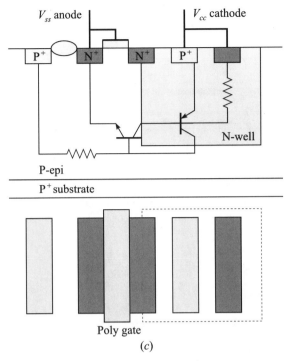

Poly gate

(c)

Figure 5-16 (*Continued*)

element when used in a voltage-controlled resistor mode [Orchard-Webb, 1991]. Another precaution with this resistor is to allow sufficient N^+-to-P^+ clearance to prevent punchthrough within the design voltage The maximum voltage capability is again the N-well-to-P-epi breakdown voltage, unless the N^+ to P^+ punches through earlier. In the pinch resistor a guard ring is also needed.

The resistance can be estimated by finding the active channel height. This can be done by subtracting the total depletion layer (both from the bottom and top) and then the resistance calculated. The depletion layer (x_d) can be calculated using the equation

$$x_d = \sqrt{\frac{2\varepsilon_0\varepsilon_{si}N_{nwell}(V_r + V_{BI})}{q}} \tag{5-8}$$

where N_{nwell} is the average N-well doping, V_{BI} is the built-in voltage, and V_r is the reverse voltage [Neudeck, 1989].

The polysilicon resistors are the only resistors with no contact to the substrate. They therefore show a high-voltage standoff. Another advantage to the isolation from the substrate is that no guard rings are required when one end is connected to an I/O pad. However, the polysilicon is thin and the resistor's current capability is limited. With silicidation the polysilicon resistivity has also

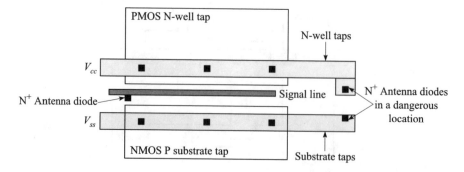

Figure 5-17. N-well and P substrate tap placements and antenna diode for I/O logic. Also shown are two antenna locations: a potential hazard for a poorly protected V_{cc}–V_{ss} supply and one that is not hazardous because of the high impedance of the signal line.

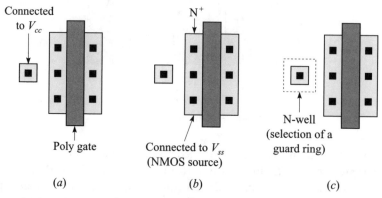

Figure 5-18. Parasitic NPN snapback may occur between two different N^+ nodes if they are sufficiently close. If the energy is dissipated in this snapback action, the result is large power dissipation over a small volume, which may result in permanent damage. (*a, b*) An N^+-to-P-epi antenna diode; (*c*) an antenna diode close to an N^+/N-well guard ring. Due to the N-well ballast resistance, the N^+/N-well structure may survive snapback, but due to its small volume, there is a thermal weak spot.

fallen dramatically. However, due to their low-capacitance and high-voltage capabilities, they are still attractive and often used.

Polysilicon resistivity is a few ohms per square; thus only resistors of few hundred ohms are possible. The length of this resistance can be large, but since the polysilicon resistors are the finest features of a chip, thin and area-efficient resistors are possible. One precaution is that the resistors should not be folded back to the high-voltage area with minimum spacing because that prevents any chance of a flash-over from the high-voltage node to the low-voltage side.

A convenient way to make extremely large resistors is by utilizing the MOS transistors. The MOS resistors have resistivity of kilohms per square and large

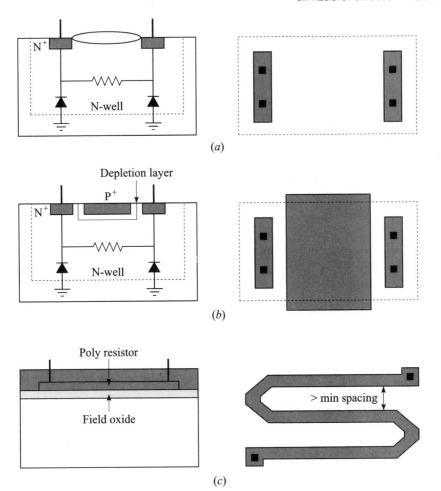

Figure 5-19. Resistor options in CMOS technology: (*a*) N-well resistor; (*b*) pinch; (*c*) polysilicon resistor.

resistors up to megaohms can be made. These are very efficient but show all the process, temperature, and voltage variations of an active circuit and fabrication process. Two methods can be implemented: One uses long channel lengths and near-full gate drives (0 V and V_{cc}) and the other utilizes reduced gate drives to scale down the length. The reduced gate drive has been discussed in Sections 2.7.2. and 2.7.3.

Sections 5.1–5.7 have dealt with individual component device layouts. The ESD pulses easily find the weakest spot in a circuit, so effort has to be made to integrate a good building block into a complete ESD safe strategy. This implementation will be discussed next.

5.8. PERIPHERY LAYOUT

Integrating the perimeter of a chip into a I/O periphery is a complex task with many factors other than ESD affecting the design. Elements that significantly influence the layout are

- ESD reliability,
- chip power delivery,
- test signals,
- clocking of I/O buffers,
- packaging options, and
- routability.

These will be discussed in the following section.

It should be noted that early planning and layout can significantly lower the layout effort. The layout is best and most reliably done using a hierarchical approach. This ensures that any cell that is properly designed, laid out, and verified can be repeatedly used with good confidence. Higher level placement, routing, and verification then becomes significantly easier. With this in mind each of the routing and power delivery options will be discussed.

5.8.1. Power Delivery

In Figure 5-20, a typical peripheral cell template is shown. These will be variants depending on the designer preference, the design requirements, and space factors. However, this is a good representation showing the peripheral and core voltage supplies. Usually the PMOS is the largest device in the periphery and will lead to a large V_{ccp} bus. The NMOS is smaller than the output device and will usually have the smaller width. Similarly, the diodes used in an ESD periphery can be tied to either the peripheral or core supplies. As seen in the layout, tying the bigger ESD diodes to the peripheral supply and the local diodes to core supplies can work well.

The peripheral bus should normally be routed on higher metals for two reasons: The higher metal is thicker so it has lower resistance and better electromigration capabilities and it lowers the capacitance between the driver device pad and the power supplies. Therefore, it reduces the overall capacitance of the output node. This in turn helps speed up the buffer.

In the previous example, the power bus (V_{cc}, V_{ss}, V_{ccp}, and V_{ssp}) was drawn as solid wires. This may be the case in general, but in some plastic packaging options this bus must be slotted [Intel, 1991, p. 30]. This slotting is essential for mechanical stress management and is highly dependent on the packaging technology rules. For example, in a plastic leadless chip carrier (PLCC) package, the shear force tends to push the wide aluminum bus toward the center of the die. This cracks the passivation over the aluminum [Gee, 1995]. Since the effect

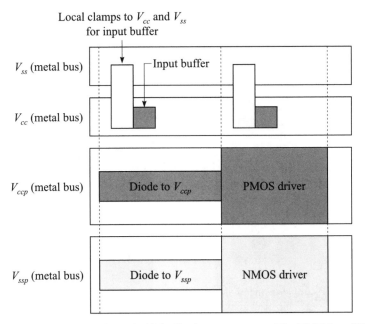

Figure 5-20. Typical input-only and I/O buffer layout strategy. The NMOS and PMOS driver sizes together with the cell pitch determine the buffer's aspect ratio. The V_{ccp} power bus should go over the PMOS and the pad clamp diode; the V_{ssp} bus should go over the NMOS and the other pad clamp diode.

peaks at the edge of the die, the I/O is usually the most affected. One solution is to place slots or contacts on a wide bus. The slotting is shown in Figure 5-21. The slotting can create major issues in power delivery because it affects current flow both directly into the core and along the peripheral supply, and it should be carefully considered early in the layout. In Figure 5-21, the metal routing from a power pad to the power bus is shown in the presence of slotting. Note that the slotting decreases the effective metal width of the power distribution both vertically and horizontally.

In general, a major concern in circuit design and reliability is the power delivery to the core. This has two aspects: electromigration reliability and DC and AC voltage quality. Since in a non-area-bumped packaging (normal wire bond, or TAB), the I/O is located at the periphery and special trade-offs are required to balance the I/O pad placement, the peripheral, and the core supplies. For example, consider Figure 5-22, which shows two power supply wires, one straight and the other with a notch. Also for such an example an electromigration (EM) limit of 1 mA/µm, a resistivity of 100 mΩ/□, a total current of 15 mA that must flow from node N_1 to node N_2, and a total path resistance of 2Ω. The straight wire meets both the electromigration and the resistance requirement, whereas the one with the notch only meets the resistance requirements. Consequently the notch can compromise the whole chip.

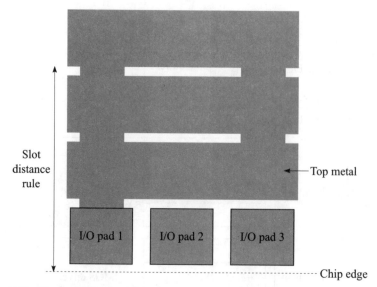

Figure 5-21. Example of metal slotting in periphery of chip to accommodate mechanical stress in some chip-packaging options.

Keeping the above example in mind, a typical power delivery strategy is examined in Figure 5-23 [Long, 1990; Annaratone, 1986]. The topmost metal, usually the thickest and with the lowest resistance, is used to deliver the power. Both ground and V_{cc} buses need to use this metal. This forces the V_{cc} and V_{ss} to cross over, as indicated [Weste, 1985]. Clearly this crossover has to be done in a lower metal (i.e., lower current density capability). The lower metal must

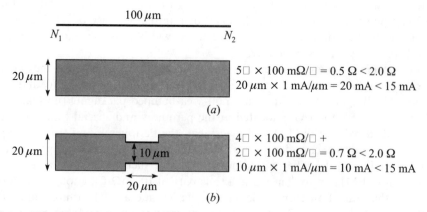

Figure 5-22. Illustration of design rules for total resistance between two points and EM requirements. The straight wire meets both the resistance and the EM requirements whereas the notched wire only meets the resistance requirements.

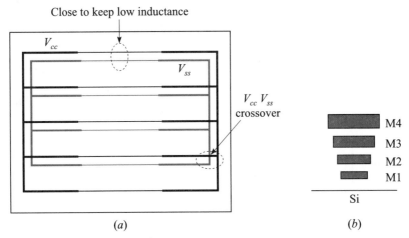

Figure 5-23. (*a*) Typical power distribution network in an IC and (*b*) metal hierarchy showing relative widths, thickness, and spacing to substrate. The power metallization is typically done using the top metal layer.

support all the current the top metal supports; otherwise, the overall electro-migration limit will be lowered. So if the original top metal is electromigration limited, then the lower metal has to be made wider than the top-layer metal. This usually complicates the layout. However, if the original top metal layer is well above the electromigration limit, then the lower limit may have the same dimension and still be robust. The width required can be calculated from the *IR* drop and the V_{cc} and V_{ss} metals must be kept close to each other to reduce the loop inductance.

Frequently, a crossover can be done in the I/O area. Here with the aid of proper power pad placement, the current can be optimized so that the cross current is minimized. The power pads should be placed close to the point of power demand such that large on-chip distribution is avoided. Then the ver-tical bars act as "equalization" bars. One such strategy is shown in Figure 5-24. The V_{ss} metal is made to cross over. The V_{ss} usually has a substrate return assisting it, whereas the V_{cc} supply is completely dependent on the peripheral power delivery. In the example, V_{cc} and V_{ss} are both in M4. The taps to the core for V_{cc} are in M4 and the taps for V_{ss} are provided in M2, after which the metal crossover is restored to M4. With the same note it should be understood that the metal devoted to power decreases the metal available for critical signal routing. This was a particular implementation of a power grid for illustrating the key concepts. A real power grid will depend on the actual package, chip, clock speed, and voltage and current requirements that may look substantially different.

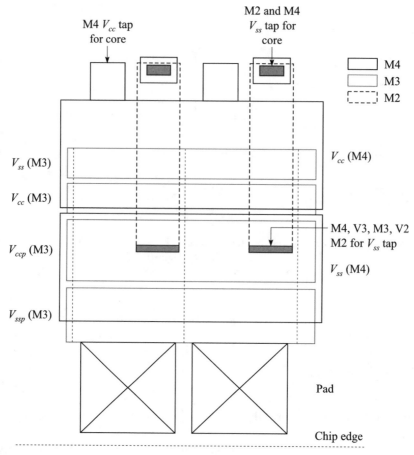

Figure 5-24. Illustration of power busing scheme showing V_{cc}-to-V_{ss} crossover.

5.8.2. I/O Routing Channels

An interesting and critical step is the allocation of routing channels for the I/O buffer. Since I/O buffer controls are usually local, only lower layer metals are required. The higher level metals are available for routing data signals back and forth to the I/O sections and any other global chip signals. This is shown in Figure 5-25.

Again early planning pays off well. Abutting layouts need planning, but they simplify and reduce routing. For example, boundary scan signals that chain the I/O buffers for testing purposes require the output of a neighbor to be the input of the next cell. The wires can be abutted on adjacent cells and valuable routing is saved. Similarly reference signals in I/O cells can have abutting layouts. Other layouts that regularly abut are the power supply,

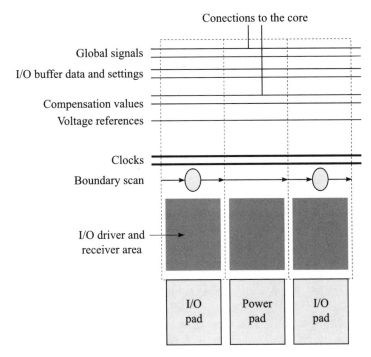

Figure 5-25. One possible I/O routing strategy for peripheral and core power supplies, local signal routing to I/O buffers and global signals.

clocks, I/O buffer mode setting options, enables, and wiring of buffer compensation values. With a well-planned cell library, even the guard rings of adjacent cells can be shared, leading to area savings. This is illustrated in Figure 5-26. Opportunities to abut should therefore be used but need careful up-front planning.

The peripheral supplies and the pad nodes can be noise generators and can have significant coupling due to large overlapping areas to wires under them. For coupling and loading reasons routing under the peripheral supplies should be evaluated carefully.

In the preceding metal width requirements due to power distribution (*IR* drop) electromigration (reliability) and signal wireability have been discussed. Another factor to consider is metal design rules due to ESD constraints.

5.9. METAL DESIGN RULES

An important design parameter is the minimum metal width required for ESD. This has two components: resistance and current density. Consider a metal wire 1500 μm long and 10 μm wide having a resistivity of $10\,\text{m}\Omega/\square$. The

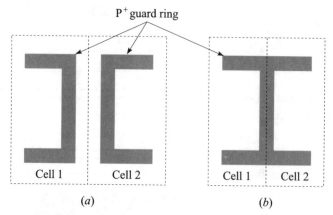

Figure 5-26. Area efficiency by using merged guard rings in abutting I/O cells.

total resistance is then 1.5 Ω. For 1 A (ESD) current the drop will be 1.5 V and the current density 100 mA/μm (across a linear edge). Assuming this is an acceptable design, can the width be scaled down by a factor of 2 if the length decreases by a factor of 2? This operation would maintain the *IR* drop. In reality, scaling down is possible until a certain minimum metal width is reached, which is dictated by electromigration or ESD limits. This ESD limit of the wiring will be considered in this section.

Early data have shown that a minimum area of 17 μm^2 was required to melt aluminum wire for a standard HBM pulse (2 kV). This was obtained by simulations by Antinone et al. [Antinone, 1986] for HBM (2 kV). The interconnect failed when the cross section was 10 μm^2. For right-angled corners a minimum width of 20 μm^2 was required to prevent melting. More recently an analytical model has been developed that can aid in the wire design [Krieger, 1991; Maloney, 1992, 1993].

One is tempted to use an *IR* drop (during an ESD pulse) to calculate the drop across a metal wire. This assumption is good as long as the metal wire is wide (low current density), but as the wire narrows, the heating effects starting becoming important. At this point the resistance of the metal increases because of the heating and in turn increases the power dissipated in the wire. Maloney has called this "bootstrap heating" [Maloney, 1992, 1993].

The adiabatic (worst case) temperature rise of the wire, *dT*, is proportional to the heat generated in it:

$$C_p(T)\, dT = I^2(t) R(T) dt \qquad (5\text{-}9)$$

and

$$C_p(T) = C_{po}(1 + \beta T) \qquad (5\text{-}10)$$

where C_{po} is the specific heat of aluminum at room temperature (0.8996 J/g-K), $\gamma = 0.0037$ for aluminum, and a measured value for AlCu is 0.0038; $\beta = 0.0005$ up to the melting point (660°C). The C_p of liquid metal is constant.

The resistance of the wire itself increases as a function of temperature and is related by

$$R(T) = R_0(1 + \gamma T) \tag{5-11}$$

and

$$R_0 = \frac{\rho_b}{W^2 \mathrm{th}^2 \rho_d} \tag{5-12}$$

where ρ_b is bulk resistivity in ohm centimeters, ρ_d is the mass density of aluminum (2.70 g/cm^3), W is metal width, and th is metal thickness.

The temperature of the wire is therefore obtained by combining the temperature rise equation and the resistance dependence on the temperature:

$$T = \left(\frac{1}{\alpha}\right)[\exp(\alpha K) - 1] \tag{5-13}$$

where

$$K = \frac{R_0}{C_{po}} \int_0^\infty i^2 \, dt \tag{5-14}$$

$$\alpha = \gamma - \beta. \tag{5-15}$$

Define

$$K_1 = \frac{\rho_\square C V^2}{2 W^2 \mathrm{th} \rho_d R_{av}} \tag{5-16}$$

and see below for other terms.

A clear consequence of this bootstrap heating is seen in Figure 5-27. For a wire with no bootstrap heating, the energy (proportional to metal sheet resistance $\rho_\square = \rho_b/\mathrm{th}$, if other factors are assumed constant) grows linearly, but as soon as bootstrap heating is factored in, there is a major increase in energy dissipated in the interconnect.

The foregoing is solved as follows: The heating is related to the I^2 term, so it is most relevant when the current density is high (i.e., thin wires or high current). Also I^2 will be the largest for CDM pulses that have the largest peak current in ESD (~ 10 A peak), and so it is usually the most damaging to the wires. Consider a CDM event. Most of the energy in the CDM event is dissipated in the arc form from the package pin to the grounding substrate. In

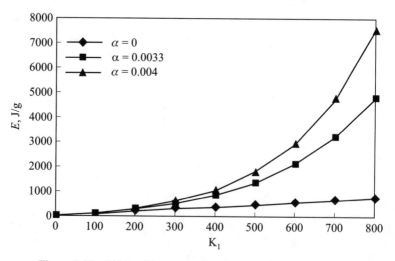

Figure 5-27. Effect of bootstrap heating as a function of α.

such a case, the energy of the discharging C at the voltage V can be equated to the dissipation in the external arc R_{av}, which is as follows:

$$R_{av} \int_0^\infty i^2 \, dt = \tfrac{1}{2} CV^2 \tag{5-17}$$

This external resistor sets the current of the discharge. The current causes energy dissipation in the wire, an this is given by the equation

$$dE = i^2(t) R_0 (1 + \xi E) \, dt \tag{5-18}$$

where

$$\xi \approx \frac{\alpha}{C_{po}} \tag{5-19}$$

Similar to the temperature equation, the energy E can be found to be

$$E = \left(\frac{1}{\xi}\right) [\exp(\xi K_1) - 1] \tag{5-20}$$

where

$$K_1 = R_0 \int_0^\infty i^2 \, dt \tag{5-21}$$

The energy required to begin melting, completely melt, and vaporize the aluminum metal is shown in Table 5-1. The width, energy, and CDM energy can be related and are shown in Figure 5-28. This clearly shows the amount of metal required for designing a desired degree of robustness. Experiments [Maloney, 1992, 1993] have shown that outright destruction by ESD requires energy considerably in excess of melting, but often less than that required for metal vaporization. The pressure required to crack the passivation is believed to be related. But any amount of melting and recrystallization implies an altered grain structure and unknown electromigration properties. Some work [Banerjee, 1996] suggests that to be safe, metal melting should be avoided. But with consideration of polymers into the interconnect scheme, the lower limit may no longer be aluminum melting (660°C) but rather the breakdown of the polymer. This is typically ~400–500°C. Thus any wiring sensitive to ESD and having polymer dielectric will have to be made wider than an oxide–metal interconnect system.

The considerations above, plus the data in Table 5-2, show that to withstand a CDM pulse an interconnect should have a cross section of ~10 μm^2. To keep

TABLE 5-1. Energy Required per Gram to Melt and Vaporize Aluminum

E J/g	Comments
$E_1 = 653$	Metal begins melting (660°C)
$E_2 = 1048$	Fully melted (latent heat of 94.5 cal/g absorbed)
$E_3 = 3172$	Metal begins to vaporize (2467°C)

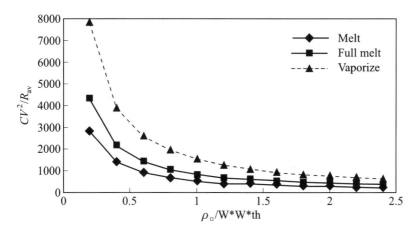

Figure 5-28. Relationship between CDM energy, metal wire physical dimensions, and its state. The CV^2/R_{av} represents the i^2 integral of Eq. 5-17.

TABLE 5-2. Maximum Capacitance Supportable by 10 μm^2 Aluminum Wire at 1500 V CDM

C (pF)	E (J/gm)	Comments
4	71.7	15% of heat for melting
11.85	653	Metal begins melting
13.98	1048	Fully melted (latent heat absorbed)
19.04	3172	Metal begins to vaporize

Source: Maloney [1992, 1993].

the resistance low, wider wires should be used. The wider the metal, the less the resistance changes and the more accurate it is to simulate using a simple circuit simulator. Another key point to remember is that if the metal is not planarized (not of uniform cross section, as can occur in sputtered metal thinning over polysilicon steps), then the minimum area at the minimum cross-sectional points should be greater than 10 μm^2.

5.10. ESD LAYOUT VERIFICATION

5.10.1. Manual ESD Verification

To a large degree, ESD verification is done manually. This task becomes increasingly difficult for two reasons:

- scaling down of the feature size and the increase in die size and
- increased number of metal and vias that make plots difficult to see.

The method to reducing mistakes is to adopt the hierarchical and modular approach suggested by Mead and Conway in the VLSI design. This hierarchical approach has worked very well for the VLSI design and should be adopted for the I/O design also. For example, consider a technology that has a pad pitch of 100 μm and a requirement of a core clamp every 5000 μm. A good modular section would be to have a module 50 pads wide. In this module all pads have ESD zap paths to any other pad. Thus, each module is designed and verified to be ESD safe by itself. Further, such modules can be enlarged and are easier to visually review on a plot. Later such modules can be abutted such that the power supplies (V_{cc}, V_{ccp}, V_{ss}, and V_{ssp}) are continuous. Since each module is ESD safe, such an assembly of modules will also result in an ESD-safe chip.

Again, within the module a standard cell approach works best. This allows the verifier to concentrate on only the interconnections between cells and not on the cell details. Thus module verification is considerably eased.

5.10.2. Automated Verification

It is the authors' experience that well-qualified cells on test chips sometimes fail in the actual chip to be protected. It is found that individual cells that are well designed are hooked up poorly and the problems lie there. This makes it imperative that an ESD layout checker be used to verify the I/O layouts against any design rule violations.

Such a tool is not easy to construct. The ESD rules can be fairly complex, especially regarding the output NMOS layout. Thus a layout checker that verifies each diffusion and via will be fairly complicated. However, this task can be simplified considerably if only the metal hookups are verified.

For this strategy, all the basic I/O, clamp, diode, and other cells are placed in a standard cell library, and the complex set of ESD rules are broken down into simpler rules, which are then validated one at a time. This library is verified manually and/or by test chips. The ESD robust cells are only allowed in this library. The library is then used to design the I/O ring based on the hookup rules. For example, it may be required to place a core clamp every 300 μm with a metal bus width of 20 μm in a particular technology. Then it becomes the task of the verifier to check that each wire between the core clamps is wider than 20 μm and the core clamps are within 300 μm. The verifying tool does not need to validate the core clamp as long it belongs to a standard ESD library. By verifying the neighboring clamp connections, the whole periphery connections are guaranteed. Each rule can then be scanned, concentrating on verifying the abutting connections only. If all the neighbors are connected right then, that rule is not violated for the whole pad ring.

5.11. SUMMARY

The layout is a crucial part of the ESD protection chain. To ensure high reliability, the following points should be particularly noted:

- The NMOS output transistor is usually the most sensitive ESD device. Special care has to be taken to ensure it faces the minimum ESD stress by channelling the current away or it snaps back very evenly.
- A properly configured ballast scheme can prevent a second breakdown in the NMOST. In unsilicided technology this was easier to implement as the diffusion was sufficient to act as the resistor. In silicided technologies a deliberate ballast can be added by the use of an N-well or silicide blocking mask.
- The ballast resistor for the NMOST works best when it is on the drain side.
- Layout of any ESD device in the current path should be extremely uniform to encourage even current flow.

- For the diodes, higher periphery-to-area ratios are preferred. The majority of the current flow is from the periphery. In some aggressively scaled-down diodes (i.e., near the breakdown even in forward bias) the inside corner of the device may get hotter than the rest of the stripe. In such a case, extra resistance (in the form of less vias) can be added to reduce the current in the corner region.

- Decoupling capacitors are essential to any VLSI circuit. These can also serve well to reduce ESD stress.

- Decoupling capacitors can be made from gate oxide (NMOS or PMOS), metal-to-metal, or polysilicon-to-metal materials. The metal-to-metal capacitors have the lowest value ($pF/\mu m^2$) but have extremely low RC time constants, whereas the gate oxide capacitors have higher values but are slower due to channel resistance.

- Decoupling capacitors must be in a format that they array easily and are area efficient.

- Resistors can be constructed in three forms: diffusion, N-well, and polysilicon. The polysilicon resistor is the only one with no diode parasitic and it has a high-voltage (but low-current) capability.

- The hook-up of the periphery I/O is important as it redistributes the ESD stress. In addition, proper planning can allow I/O areas to be used for global routing, leading to better chip layouts.

- Outright destruction of metal by CDM requires heating beyond melting, approaching the vaporization energy. However, avoidance of all metal melting is desirable. The CDM pulse requires a metal cross section of $\sim 10\,\mu m^2$ to withstand a 1-kV zap with minimal disturbance of the metal. The heat generated by the passage of the current increases the resistance of the wire, which in turn heats more with the passage of more current. This is the bootstrap heating effect. Wider metals not only decrease the resistance but also reduce the resistance buildup during a CDM pulse due to the bootstrap heating effect of temperature.

REFERENCES

[Amerasekera, 1995] A. Amerasekera and C. Duvvury, *ESD in Silicon Integrated Circuits*, Wiley, Chichester, England, 1995.

[Annaratone, 1986] M. Annaratone, *Digital CMOS Circuit Design*, Kluwer, Boston, 1986, p. 316.

[Antinone, 1986] R. J. Antinone, et al., *Electrical Overstress Protection for Electronic Devices*, Noyes Park Ridge, NJ, 1986, p. 52.

[Baker, 1989a] L. Baker, R. Currence, S. Law, M. Lee, C. Lee, S. T. Lin, and M. Teened, "A Waffle Layout Technique Strengthens the ESD Hardness of the NMOS Output Transistor," *Proc. EOS/ESD Symp.*, Sept. 1989, p. 175.

[Baker, 1989b] F. K. Baker and J. R. Pfiester, "The Influence of Tilted Source-Drain Implants on High-Field Effects in Submicrometer MOSFETs," *IEEE Trans. ED.*, **35**(12), 1988, p. 2119.

[Banerjee, 1996] K. Banerjee, et al., "Characterization of VLSI Circuit Interconnect Heating and Failure under ESD Conditions," *Int. Reliability Symp. Proc.*, 1996, p. 237.

[Carbajal, 1992] B. G. Carbajal III, R. A. Cline, and B. H. Andresen, "A Successful HBM ESD Protection Circuit for Microns and Sub-Micron Level CMOS," *Proc. EOS/ESD Symp.*, 1992, p. 234.

[Christou, 1994] A. Christou, Ed. *Electromigration and Electronic Device Degradation*, Wiley, New York, 1994.

[Diaz, 1994] C. Diaz and G. Motley, "Bi-modal Triggering for LVSCR ESD Protection Devices," *Proc. EOS/ESD Symp.*, 1994, p. 106.

[Duvvury, 1991] C. Duvvury and R. Rountree, "A Synthesis of ESD Input Protection Scheme," *Proc. EOS/ESD Symp.*, 1991, p. 88.

[Duvvury, 1995] C. Duvvury and A. Amerasekera, "Advanced CMOS Protection Device Trigger Mechanism During CDM," *Proc. EOS/ESD*, 1995, p. 162.

[Gee, 1995] S. A. Gee, M. R. Johnson, and K. L. Chen, "A Test Chip Design for Detecting Thin Film Cracking in Integrated Circuits," *IEEE CHMT-B*, **18**(8), 1995, p. 478.

[Hulett, 1981] T. V. Hulett, "On Chip Protection of High Density NMOS Devices," *Proc. EOS/ESD Symp.*, 1981, p. 90.

[Intel, 1991] Intel Corp., "Components Quality and Reliability," Chapter 8, 1991.

[Jeng, 1994] S. P. Jeng, M.-C. Chang, T. Kroger, P. McAnally, and R. H. Havemann, "A Planarized Multilevel Interconnect Scheme with Embedded Low-Dielectric Constant Polymers for Sub-Quarter Micron Applications," *1994 Symp. on VLSI Tech.*, 1994, p. 73.

[Ker, 1992] M. D. Ker, C. Y. Wu, and C. Y. Lee, "A Novel ESD/EOS Protection Circuit with Full-SCR Structures," *Proc. EOS/ESD*, 1992, p. 258.

[Ker, 1996] M. D. Ker, C. Y. Wu, T. Cheng, and H. H. Chang, "Capacitor-Couple ESD Protection Circuit for Deep-Submicron Low-Voltage CMOS ASIC," *IEEE Trans. VLSI Syst.*, **4**(3), 1996, p. 307.

[Krakauer, 1992] D. Krakauer, and K. Mistry, "ESD Protection in a 3.3V Sub-Micron Silicided CMOS Technology," *Proc. EOS/ESD Symp.*, 1992, p. 250.

[Krieger, 1991] G. Krieger, "Non-uniform ESD Current Distribution Due to Improper Metal Routing," *Proc. EOS/ESD Symp.*, 1991, p. 104.

[Larson, 1997] P. Larson, "Parasitic Resistance in an MOS Transistor Used as On-Chip Decoupling Capacitance," *IEEE J. Solid State Circuits*, **32**(4), 1997, p. 574.

[Lee, 1997] T. C. Lee and J. Cong, "The New Line in IC Design," *IEEE Spect.*, March 1997, p. 52.

[Long, 1990] S. I. Long and S. E. Butler, *Gallium Arsenide Digital Integrated Circuit Design*, McGraw-Hill, New York, 1990, p. 266.

[Maene, 1992] N. Maene, J. Vandenbroeck, and L. V. D. Bempt, "On Chip Electrostatic Discharge Protections for Input, Outputs and Power Supplies of CMOS Circuits," *Proc. EOS/ESD Symp.*, 1992, p. 228.

[Maloney, 1988] T. J. Maloney, "Designing MOS Inputs and Outputs to Avoid Oxide Failure in the Charged Device Model," *Proc. EOS/ESD Conf.*, 1988, p. 220.

[Maloney, 1992] T. J. Maloney, "Integrated Circuit Metal in the Charged Device Model: Bootstrap Heating, Melt Damage, and Scaling Laws," *Proc. EOS/ESD Symp.*, 1992, p. 129.

[Maloney, 1993] T. J. Maloney, "Integrated Circuit Metal in the Charge Device Model: Bootstrap Heating, Melt Damage, and Scaling Laws," *J. Electrost.*, **31**, 1993, p. 313.

[Millman, 1987] J. Millman and A. Grabel, *Microelectronics*, McGraw-Hill, New York, 1987, p. 197.

[Miyamoto, 1997] M. Miyamoto and T. Takeda, "High-Speed and Low-Power Interconnect Technology for Sub-Quarter-Micron ASIC's" *IEEE Trans. ED*, **44**(2), 1997, p. 250.

[Neudeck, 1989] G. W. Neudeck, *The PN Junction Diode*, Addison-Wesley, Reading, MA, 1989, Chapter 3, p. 33.

[Orchard-Webb, 1991] J. H. Orchard-Webb, "A Characterization of Components for an Optimized CMOS Input Protection System," *Proc. EOS/ESD Symp.*, 1991, p. 83.

[Paraszczak, 1993] J. Parazczak, D. Edelstein, S. Cohen, E. Babich, and J. Hummel, "High Performance Dielectrics and Processes for ULSI Interconnection Technologies," *Proc. IEDM*, 1993, p. 261.

[Polgreen, 1992] T. L. Polgreen and A. Chatterjee, "Improving the ESD Failure Threshold of Silicided NMOS Output Transistor by Ensuring Uniform Current Flow," *IEEE Trans. ED*, **39**(2), 1992, p. 379.

[Ramaswamy, 1995] S. Ramaswamy, P. Raha, E. Rosenbaum, and S. M. Kang, "EOS/ESD Protection Design for Deep Submicron SOI Technology," *Proc. EOS/ESD Symp.*, 1995, p. 212.

[Ramaswamy, 1996] S. Ramaswamy, C. Duvvury, A. Amerasekera, V. Reddy, and S. M. Kang, "EOS/ESD Analysis of High-Density Logic Chips," *Proc. EOS/ESD Symp.*, 1996, p. 285.

[Razavi, 1995] B. Razavi, K. F. Lee, and R. H. Yan, "Design of High-Speed, Low-Power Frequency Dividers and Phase-Locked Loops in Deep Submicron CMOS," *Proc. IEEE JSSC*, **30**(2), 1995, p. 101.

[Strauss, 1991] M. S. Strauss and M. H. White, "Protecting N-Channel Output Transistors from ESD Damage," *Proc. EOS/ESD Symp.*, 1991, p. 110.

[Stricker, 1995] A. Stricker, D. Gloor, and W. Fichtner, "Layout Optimization of an ESD-Protection n-MOSFET by Simulation and Measurement," *Proc. EOS/ESD Symp.*, 1995, p. 205.

[Tong, 1996] M. Tong, R. Gauthier, and V. Gross, "Study of Gated PNP as an ESD Protection Device for Mixed-Voltage and Hot-Pluggable Circuit Applications," *Proc. EOS/ESD Symp.*, 1996, p. 280.

[Voldman, 1992] S. H. Voldman, V. P. Gross, M. J. Hargrove, J. M. Never, J. A. Slinkman, M. O. O'Boyle, T. M. Scott, and J. Deleckl, "Shallow Trench Isolation Double-Diode Electrostatic Discharge Circuit and Interaction with DRAM Output Circuitry," *Proc. EOS/ESD Symp.*, 1992, p. 277.

[Voldman, 1994a] S. H. Voldman, "ESD Protection in a Mixed Voltage Interface and Multi-Rail Disconnected Power Grid Environment in 0.5- and 0.25-μm Channel Length CMOS Technologies," *Proc. EOS/ESD Symp.*, 1994, p. 125.

[Voldman, 1994b] S. H. Voldman, S. S. Furkay, and J. R. Slinkman, "Three-Dimensional Transient Electrothermal Simulation of Electrostatic Discharge Protection Circuits," *Proc. EOS/ESD Symp.*, 1994, p. 246.

[Wada, 1995] T. Wada, "Study of Evaluation Methods for Charged Device Model," *Proc. EOS/ESD Symp.*, 1995, p. 186.

[Wei, 1993] Y. Wei, Y. Loh, C. Wang, and C. Hu, "Effect of Substrate Contact on ESD Failure of Advanced CMOS Integrated Circuits," *Proc. EOS/ESD Symp.*, 1993, p. 221.

[Weste, 1985] N. H. E. Weste and K. Eshraghian, *Principles of CMOS VLSI Design A Systems Perspective*, Addison-Wesley, Reading MA, 1985, p. 226.

CHAPTER 6

ESD AND I/O INTERACTIONS

Until this point ESD and I/O design have been discussed in separate chapters and as separate topics. However, in reality they are interdependent and these dependencies will be examined here. To initiate this examination, a brief overview of chip performance and its link to the I/O and ESD design cycle is a good starting point.

The chip's target performance sets the I/O bandwidth requirement, and this in combination with several other factors (such as power, compatibility, topology, speed, noise margins, etc.) leads to the selection of a signaling scheme. The signaling scheme is in turn mapped to I/O buffers. For optimum performance the I/O buffer design should be allowed the entire design space that the logic design enjoys. However, reliability issues such as ESD and the hot electron impose certain constraints in I/O design and take away flexibility. The ESD reliability as an added constraint ends up reducing the number of options available to an I/O designer. The key objective is to minimize these constraints and maximize the design space.

The I/O and ESD interaction falls into three main categories:

- *Speed, Area, and ESD Performance Trade-off for I/O Buffer.* These are caused by the granularity in width and fixed lengths imposed by some ESD Rules.
- *Coupling of Periphery Noise to Core Power Supply.* This occurs when there is sufficient I/O noise generated by the I/O switching and allows the ESD diodes to couple enough noise into the core power supply.
- *ESD Diode Termination Effects in I/O Signal Integrity Due to Large Under- and Overshoots of I/O Voltage.* The large steps are usually caused

by a conjunction of topology, lack of buffer compensation, and aggress-ive timing requirements.

In this chapter interactions between I/O and ESD requirements, its conse-quences on I/O, and the chip will be considered.

6.1. I/O PERFORMANCE TRADE-OFF

The I/O performance trade-off can be categorized into limitations imposed on output buffer sizing, compensation scheme tracking, and input issues due to mixed-voltage compatibility.

6.1.1. Impact on Output Buffer Sizing Granularity

In some technologies, allowable finger widths (or ranges) are experimentally determined. these widths are then frozen and are used in all subsequent designs with no variations permitted. This can be a major constraint imposed by ESD on the output buffer sizing. Typically the buffers characterized include stan-dard sizes (e.g., 32 mA or 40-Ω buffers) commonly encountered or projected for the technology being designed. However, I/O may require much finer gran-ularity when custom I/O and signaling are desired. For example, an NMOS driver of 200/0.5 may be required. If the minimum ESD finger width allow-able is 40/0.5 µm, five legs will satisfy this requirement. However, if the ESD design rule required a width of 60/0.5 µm, the choices for the 200/0.5 driver are 180/0.5 or 240/0.5, which are both 10% under or 20% over the required drive strength. The reduced or excess drive strength has bearings on network settling and noise generation. This width issue can usually be compensated for by increasing the gate length, resulting in the required drive strength but with larger gate capacitance and a corresponding increase in predriver size (i.e., increased area and power) and an additional delay in signal propagation through the driver. In some cases, the ESD rules may be so rigid that even changes in device length are not permitted. This can effectively lock in the granularity in the I/O design with its corresponding penalties. In the area of high-speed custom I/O, this granularity can be a significant disadvantage.

If more flexibility is sought, then the NMOS output device should be made independent of ESD requirements. Here the N-well resistor current saturation effect can be employed to safeguard the NMOS device. As explained in Chapter 2, the second breakdown current limit of a finger, which depends on the finger width, has to be greater than the N-well saturation current. Thus based on N-well geometry, a minimum finger size is chosen. For example, consider a saturation current for the N-well of $\sim$1–2 mA/µm and an I_{t2} of $\sim$4–5 mA/µm; then if the minimum N-well size is 4µm wide, the minimum NMOS size will be 2 µm. Larger NMOS gate sizes are permitted. In addition, no constraints on the channel length are imposed, though in some cases hot-electron requirements may require them. They provide sufficiently fine granularity, and that is a merit of the saturated N-well resistance technique.

The PMOS device does not snap back and destroy itself as easily as the NMOS; therefore, more layout freedom exists. However, a subtle factor in the PMOS layout strategy shown in Figure 5-5 exists. Here the drain is shared with a diode. As the PMOS device scales up, the diode size also increases. This diode may be far above the ESD requirements, and it is essentially an excess capacitive load at the I/O pad.

The PMOS can scale up because the output device may have to be cascaded, in which case the size approximately doubles or the channel length increases. These will be discussed in the next chapter. Another reason for a larger PMOS driver is that a dual-mode driver may be required. In a dual-mode driver, a high- and a low-voltage mode may be supported by a driver. For example, consider a 5-V technology that needs to support a 5-V and a 3.3-V CMOS driver. Also assume that the threshold voltages for the PMOS and the NMOS are 1 V each. Therefore, to match the 3.3-V driver saturation current, the driver needs to scale up by approximately $(5 - 1)^2/(3.3 - 1)^2$, or 3 times larger. If the linear current and the output resistance have to be matched, then the increase is $(5 - 1)/(3.3 - 1)$, or 1.7 times the 5-V driver size. In the latter case, if it is assumed that the 5-V driver had the right-size diode, then the 3.3-V driver will have an extra 70% diode, which is not essential, and it acts as an excess capacitance at the I/O pad. Alternatively, if the PMOS size scales down, then the minimum diode size requirement forces the PMOS device to be sized back up and the extra legs to be turned off.

6.1.2. Impact of Compensation Scheme Implementation

The ESD can impose constraints in the compensation of I/O buffers. For the digital style of compensation, clearly the minimum granularity is set by the minimum finger width, as discussed in Section 6.1.1. If the granularity is large, this will cause large jumps in the I/O drive capability of the buffer, leading to poor compensation features. Also some schemes such as the binary weighting and the equal percentage change weighting may not be viable or only poorly implemented.

For analog compensation the buffers pose a different problem, a problem of difference in the bias generation transistor and the point-of-use transistor. Consider, for example, the scheme in Figure 6-1. There can be two differences between the compensation generation and the point of use:

- The N-well resistor is needed for the compensation generation transistor for its ESD protection but is not required at the point-of-use transistor stack. This causes small variations in compensation generation and its use.

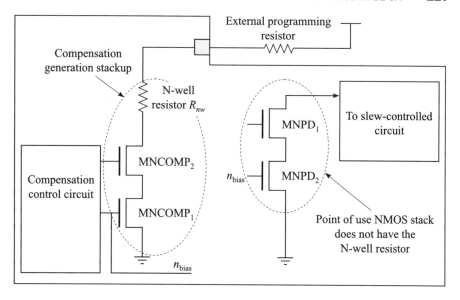

Figure 6-1. The ESD design rules can result in differences in bias generation and application at the point of use reducing the effectiveness of the compensation.

- The output NMOS driver gate lengths for the calibration transistors may not be the same as the point-of-use stack. The calibration stack channel length follows an ESD guide line that may be flexible or not. The pre-driver transistor $MNPD_2$ usually has a large gate length transistor to minimize the drain modulation effects. Therefore another source of bias generation and use arises.
- The stacking of the NMOS gate leads to a different ESD tolerance. The stacked gates may be required in some compensation schemes. Tong [1996] shows that between two NMOSs (single and cascaded), both protected by N-well resistors, the ESD single-gate performance was 2.9 kV, which dropped to 2.3 kV for the cascaded transistors. This performance was recovered as soon as a parallel path to V_{dd} was installed. The tolerance of the single NMOS improved to 6.6 kV, whereas the cascaded-gate tolerance improved to 7.6 kV.

These issues are not insurmountable but require additional circuit design and may entail delay, some error in compensation, and area or power penalties.

Another subtle effect imposed by ESD design on I/O and testability is to prevent easy access of nodes inside a circuit. To access an internal node and bring it out as a pin, an I/O structure has to be interposed, which in turn hinders the circuit operation to be monitored.

6.1.3. Input Pass Transistor Jitters

Pass transistors are employed to protect input circuits when there is a high-voltage incoming signal (say 5 V) to a chip that is fabricated in a low-voltage technology (say 3.3 V) [Wong, 1988] Such circuits are shown in Figure 6-2. The function of the pass transistor is to limit the voltage at the input node B. The gate of the pass transistor is tied to V_{cc}, so that the drain and gate are never stressed. Therefore node B can be pulled up to a maximum of $V_{cc} - V_{tn}$.

How the pass transistor is tied up can play an important role in its ESD protection and a weaker role in performance. For this, consider the example shown in Figure 6-2, when three types of pull-up techniques for the pass gate are shown. One option is a direct pull up to V_{cc}, the second through a PMOS, and the third through a resistor. It should be noted that there is also the capacitive coupling that goes through the gate–drain overlap into the gate of the pass transistor, which together with the tie-up option from an RC circuit. This is very similar to the GCMOS operation, only the RC effect has to be minimized in this case during normal operation.

Two effects are seen in terms of performance:

- The first is that the input node no longer sees a balanced waveform (i.e., the voltage is clipped to $V_{cc} - V_{tn}$ at the high input, although the input voltage is higher). this can be seen in Figure 6-2. This clipping may require retuning of the input receiver to the reduced swing input, and it also leads to reduced noise margin at the input. The pass transistor also introduced delay in the input path.

- The gate charging and discharging influence the NMOS performance, speeding it up in one direction and slowing it down in the other. Consider a fast-rising edge applied to the input that couples charge into the gate of the pass transistor. If the RC time constant is comparable or larger than the rising edge rate, then there is a higher voltage built up at the gate, which in turn enhances the conduction through the pass transistor. If the RC time constant is much smaller than the input edge rate, no significant speed-up occurs as the gate rapidly discharges through the resistor. For the falling edge, the opposite holds, as illustrated in Figure 6-2. Therefore, poorly tied gates can lead to signal skew between rising and falling edges as well as delays in the falling edge. These uncertainties will reduce the timing budget. With respect to delays and skews the direct gate tie-up performs the best and the PMOS pull-up the worst. By proper choice of the pass transistor and the PMOS, this can be easily adjusted to tolerable limits. The trade-offs are shown in Table 6-1.

However, the same RC effect can be used to encourage current to flow in other (designed) paths during an ESD event. By placing a high impedance as a tie-up between the gate and the power supply, the ESD discharge current through the gate is limited; therefore the robustness increased. For a well-

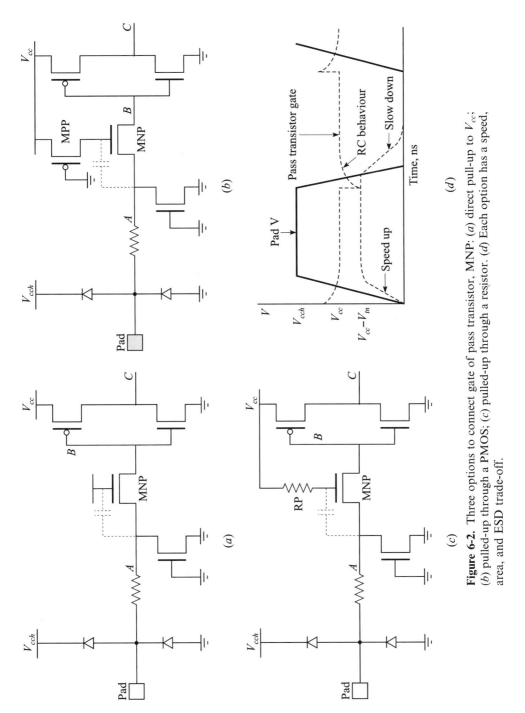

Figure 6-2. Three options to connect gate of pass transistor, MNP: (*a*) direct pull-up to V_{cc}; (*b*) pulled-up through a PMOS; (*c*) pulled-up through a resistor. (*d*) Each option has a speed, area, and ESD trade-off.

TABLE 6-1. Summary of Speed and ESD Performance Trade-off

Pass Gate Tie-up Option	ESD Performance	Speed Performance	Area
V_{cc}	Poorest	Best	Smallest
PMOS pull-up transistor	Best	Poorest	Medium
Pull-up resistor	Good	Good	Largest

protected circuit, this additional protection is small (maybe about 100 V CDM), but it is much more pronounced if the overall chip protection is poor.

If the pass gate issue needs to be minimized, two solutions can be adopted. One is to add a significant capacitance to the pass transistor gate (maybe 10 times the drain-to-gate overlap capacitance). This will minimize any large variations in the gate voltage due to fast edges. The other is to use a large PMOS as a tie-up. Only this PMOS is on when the circuit is operational, thereby reducing the RC time constant well below the input edge rate. When the chip is not powered up, the PMOS is turned off and therefore it is an open circuit. This will give the best ESD performance. This scheme is illustrated in Figure 6-3.

6.1.4. Other Interactions

The following topics have been discussed earlier in the text. They are briefly mentioned here again to remind readers of the ESD and I/O interaction possibility in those particular areas:

- The length of the I/O driver needs to be sufficiently large such that the ESD protection clamp triggers on before the output driver (especially the NMOS). The snapback voltage is a function of the channel length, and observed values suggest an $L^{0.3}$-to-L dependence. Such scenarios were presented in Section 2.8. The larger I/O driver has a higher capacitance therefore loading the bus. This in turn degrades the edge rate.

- Similarly the predriver lengths need to ensure that they do not snap back sooner than the clamps, thereby allowing a high voltage to directly appear on the I/O buffers. This is mainly a concern for poorly protected circuits.

- If a RGNMOS or a GCNMOS scheme is implemented to encourage uniform ESD current flow in NMOS transistors, then the predrivers need to be sized appropriately. In such cases an even number of predriver inverting stages should be tied to the V_{ccp} supply. This was discussed in Section 3.5.

- The decoupling capacitor required for I/O operations directly helps ESD performance. Consequently larger decoupling capacitors are helpful to the ESD and generally for the I/O. The decoupling capacitor effects are discussed in Section 3.1.

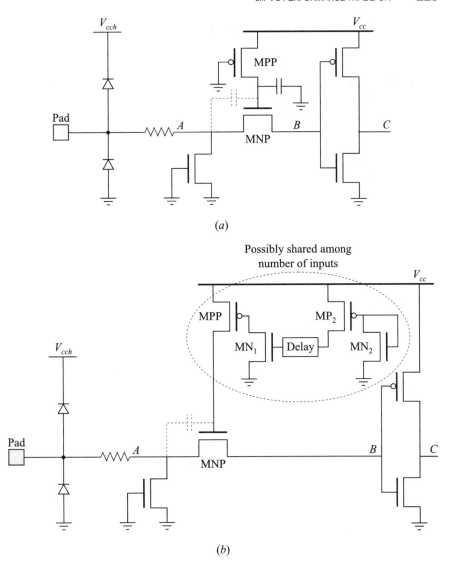

Figure 6-3. Two options to rectify pass gate issue: (*a*) add a capacitance at the pass transistor gate, thereby drastically reducing the voltage developed; (*b*) add a large PMOS device that is off when the circuit is not in operation and on when the circuit is in operation.

6.2. PERIPHERY NOISE COUPLING INTO CORE SUPPLIES

During I/O switching, noise is generated due to the instantaneous current switching through the power supply inductance. The peripheral noise can be injected into the core power supply by two means: shared board and package parasitic paths and coupling due to ESD structures. The shared board and package paths are shown in Figure 6-4. These are inherent in the package and board design. Here ESD contribution to noise coupling will be discussed.

In the methodology discussed in Chapter 3, the periphery and core are coupled using diodes for ESD protection. It can be argued that it is possible to avoid this coupling by use of cantilevered diodes as power clamps on the periphery, thereby unlinking the periphery and core supplies. However, there is

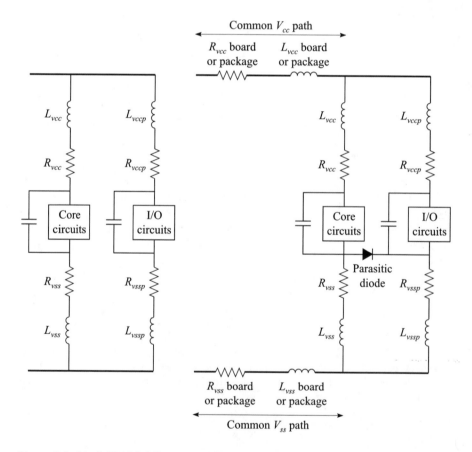

Figure 6-4. Ideal CMOS I/O power delivery system compared to real CMOS power delivery system. In the ideal power delivery system no common path exists between the I/O and the core circuits; therefore, no noise is coupled. The common path for the I/O and the core supplies couples noise into the other power supply.

always one parasitic diode between the V_{ssp} and V_{ss} that is inherent in all bulk P-epi CMOS processes. This can couple noise into the core V_{ss} under certain noisy circumstances, but it is usually a weak influence. When there are deliberate coupling diodes in addition to the parasitic diodes, larger coupling is possible. Generally a stronger noise injector into the core supply is the clamping action of the ESD diode when an input pad voltage swings far in excess of V_{cc} or much below V_{ss}.

6.2.1. I/O and Core Switching Current Characteristics

Before discussing the ESD contribution to noise, it is worthwhile to comprehend the current requirements of the I/O and core circuits. The core circuits switch near the core clock edge and have sharp transitions. The I/O circuits switch a little later, as the data goes out from a latch to the predriver and then to the driver. Also, the I/O drivers and predrivers are adjusted to have controlled slew rate switching. Thus the slew rate of the I/O and core circuits may easily be a factor of 2–4× slower. Figure 6-5 shows the timing-and-magnitude relationship between the core and the I/O current requirements for logic-oriented (e.g., microprocessor) circuits. For memory circuits, the I/O may be the significant current consumer on the chip compared to the core circuits. Traditionally, in both the memory and the microprocessors, the peripheral supplies have been termed "dirty" because they are very noisy; consequently separate dedicated pins have been assigned for their power supply in an effort not to "contaminate" the core supplies [Hanafi, 1992]. According to Rent's rule, the number of core circuits increases much faster than the I/O buffers required to support these circuits. Therefore in the future the ratio of core currents to I/O currents should be expected to increase further.

Partitioning the periphery and core supplies divides up the pins and the power planes in a package. However, it is seen that the peak I/O and core currents do not switch simultaneously. This raises the question of whether the I/O and ESD pins and package can be shared or separate power distribution is essential for both. Combining the pins and package planes can reduce the di/dt noise for both the core and the I/O circuits and also reduce ESD issues. However, each option needs to be analyzed in detail to verify such assumptions.

Usually circuits have employed a separate "clean" and "dirty" supply strategy. In the ideal CMOS power distribution shown in Figure 6-4 there is no common path between the I/O and core supplies except at the ideal power supply. There are no coupling diodes, and this ideal CMOS power system is truly independent and no cross coupling is possible between core and I/O power supplies. Thus the I/O noise generated will not affect the core operation.

However, in real life the parasitic diode between V_{ssp} and V_{ss} may couple the periphery to the core as shown in Figure 6-4. If the periphery is very noisy, then the diode can be forward biased. Further, there is eventually a common path for both the I/O and core currents, which will inject a common mode noise into

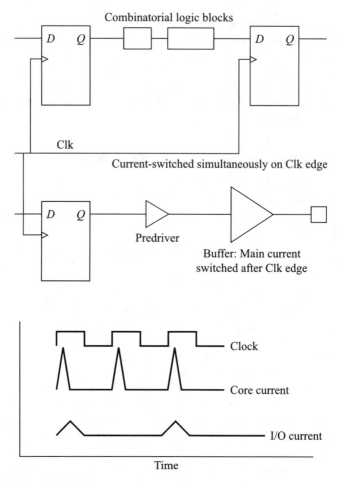

Figure 6-5. Typical I/O and core current requirements differ in time, switching speed, frequency, and amplitude. These should be taken into account when partitioning the power planes of a package.

the core power supply. This common path is shown as the packaging and the board parasitic and these will be the paths where noise is coupled.

6.2.2. Peripheral Noise Coupling Through ESD Diodes

Consider the behavior of an I/O buffer switching a large capacitor, as shown in Figure 6-6. If the system is *RC* dominated as would happen for a slow system, then the parasitic diode will forward bias when the noise voltage exceeds the forward-bias voltage of the diode. It should be noted that since the chip may be hot (100°C) and the diode large (a diode at each pad), the voltage required to initiate coupling in some circumstances may not be high. The peak current

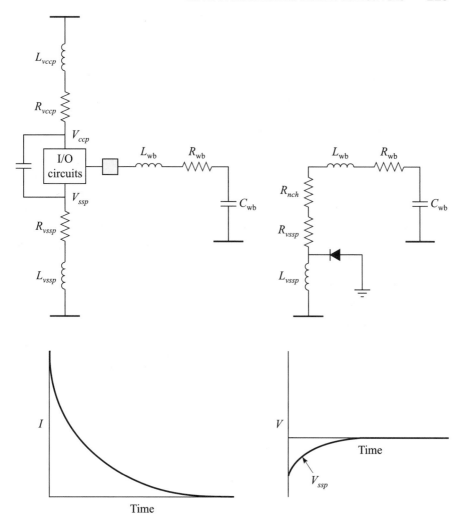

Figure 6-6. Simplified CMOS driver circuit approximated as dominantly RC and amenable to analytical solutions. The transistor is assumed to switch on rapidly compared to the RC time constant.

is approximately set by the initial DC voltage and the resistance of the circuit. The peak rate of current decay is $1/RC$. Therefore the maximum negative $L \, di/dt$ noise seen at the V_{ssp} node is at time zero and decays from there. When the V_{ssp} noise is lower than the diode cut-in voltage $V_d (V_{ssc} - V_{ssp} > V_d)$, the parasitic diode conducts. Since the diode is a non-linear, exact analysis is no longer simple. The onset of conduction is shown in Eq. 6-1. In slower I/O systems, corresponding to higher voltages, V, this noise coupling was a possibility. With the same token, since the core supply was high, this coupled noise was only a small fraction:

$$\frac{L_{vssp}}{(R_{vssp} + R_{wb} + R_{nch})^2 C} \leq \left(\frac{V_d}{V}\right) \tag{6-1}$$

where L_{vssp} and R_{vssp} are the parasitic inductance and resistance of the periph-eral supply, R_{wb} is the resistance of the wire bond (extremely small), and R_{nch} is the output resistance of the NMOS device assumed to be linear and constant.

It should be noted that since the power supply voltages are decreasing rapidly (e.g., 5 V, 3.3 V, 1.8 V, 1.3 V, 1 V, ...), the possibility of injecting excess noise (> 0.4 V) also decreases. For a 1-V system a typical allowable noise target may be 10%, which is too low to initiate diode conduction. This is very similar to why BiCMOS circuits are not preferred in low-voltage technologies since the BJT turn-on voltage is a significant part of the power supply voltage. However, for the older 5-V technologies this type of noise injection can still occur.

For faster I/O no longer RC dominated, LRC analysis is necessary [Dabral, 1993; Gabara, 1996, 1997]. If the circuit is dominantly LC (rare), similar analy-sis can be done to predict the maximum negative di/dt and the conditions for noise coupling can be calculated. However, it is more realistic to examine LRC circuits. For highly underdamped LRC circuits the current waveform is illus-trated in Figure 6-7, Eq. 6-2, and Eq. 6-3, which show the noise coupling

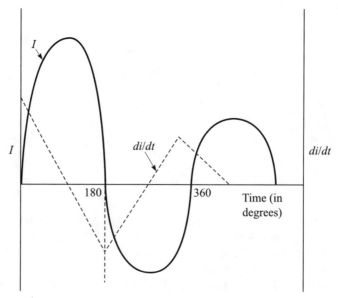

Figure 6-7. Current waveforms for an LC dominated I/O system. The maximum noise at 180°.

criteria for an LRC circuit. Here again the starting DC voltage plays a key role in determining the maximum current and consequently the di/dt noise. In future systems the DC voltage will scale down, so contributions due to V will decrease.

The instantaneous current forms an exponentially decaying sinusoid, as expressed in the following equations with frequency ω and damping factor α:

$$i(t) = \frac{V}{(L_{vssp} + L_{wb})} e^{-\alpha t} \sin \omega t$$

$$\omega = \sqrt{\omega_0^2 - \alpha^2}$$

$$\alpha = \frac{R}{2(L_{vssp} + L_{wb})} \tag{6-2}$$

$$\omega_0 = \frac{1}{\sqrt{LC}}$$

when $\alpha \ll \omega_0$, and setting the maximum negative-going noise to π radians, the noise required to initiate diode coupling can be simplified to

$$\left(\frac{L_{vvsp}C}{(L_{vvsp} + L_{wb})C}\right) \exp\left[-\left(\left(\frac{R}{L}\sqrt{LC}\right)\frac{\pi}{2}\right)\right] < \left(\frac{V_d}{V}\right) \tag{6-3}$$

There are several methods that reduce ground noise in addition to slew rate control, discussed in Chapter 4:

- increase the number of V_{ssp} pins (V_{ccp}),
- improve the package power planes, and
- provide decoupling capacitors at all levels in the chip package and board.

Once the diodes turn on, analytical modeling is very difficult and simulations have to be restored. The noise injected from the peripheral power supply into the core power supply is considered next. It is also compared with the effect of addition of ESD diodes to cross couple the supplies.

6.2.3. Peripheral Noise and Power System Impedance Interaction

With the onset of diode coupling, circuit simulation is necessary. The following qualitatively describes the coupled noise between the periphery and core due to the ESD components utilizing one such circuit simulation setup. First a simple I/O and power model is set up. In this model there are a variable number of coupling diodes between V_{ssp} and V_{ss} and between V_{ccp} and V_{cc} used to couple the power supplies for ESD reasons. These can be added one at a time, except for the parasitic diode between V_{ssp} and V_{ss}, which always exists. There are decoupling capacitors between the core supplies and the periphery supplies.

The noise source in the periphery is the I/O buffer switching simultaneously and discharging a capacitor and having some crowbar current. This noise is coupled into the core. Two extremes of core power supply impedance are shown, one with high impedance and the other with low impedance.

The peak noise in the core supply (deviation from nominal $V_{cc} - V_{ss}$) is shown in Figure 6-8. This case can represent a no-substrate tap option or an option where pins have deliberately been conserved. When a diode is added, the noise in the core decreases. This happens despite more coupling from the diodes because the diodes not only couple noise into the core but help dissipate this noise through a lower impedance path. In this case the net AC impedance seen by the core (say the V_{ssc} node) is reduced by addition of the parallel V_{ssp} and V_{ccp} paths. In this case the benefit of adding the parallel net is greater than the noise coupling due to the diodes. With addition of more diodes in the chain, the effectiveness of the coupling diode diminishes and the noise performance is similar to the no-coupling diode case.

Similarly, if diode resistance is added, it effectively reduces the injected noise into the core. However, the increased resistance also decouples the core and peripheral supplies and consequently is no longer an effective parallel path. This can be seen in Figure 6-8 where the diode coupling with no resistance shows lower noise as opposed to the case with diode resistance.

If a substrate tap is added to the scheme (or more power pins are assigned), as was really intended in this package design, the noise levels reduce significantly. This is shown in Figure 6-9. With no coupling diode the injected noise in the core is small. However, with the addition of one diode, the noise increases drastically. This can be accounted for primarily by additional noise injection and only a marginal decrease in the power supply impedance as seen by the core. The effective impedance of the inductors was calculated to be ~ 0.1–$1\,\Omega$ (at 100 MHz). The substrate tap is typically tens of milliohms. Thus by coupling more parallel paths (~ 0.1–$1\,\Omega$) to an already low-impedance path, it does not reduce the impedance but allows more noise coupling. This is seen as an effective increase in the core power supply noise. As more diodes are added, the coupling is reduced, and by addition of four diodes, the noise levels are restored to the no-diode case.

The diode resistance also acts to suppress the coupling. This is observed when the data with and without diode are compared. For the case of one diode with resistance the core noise levels are lower as opposed to the case with no diode resistance.

When placing coupling diodes, the complete chip power supply impedance should be considered. If the core has a low-impedance core power supply (generally the case), the addition of one coupling diode may inject more noise rather than help reduce the power supply impedance. Thus several cascaded diodes should be employed. However, if the core power supply impedance is high, addition of one diode between the peripheral and core supply can effectively lower the power supply impedance of the core and smaller diode chains can be employed [Dabral, 1994].

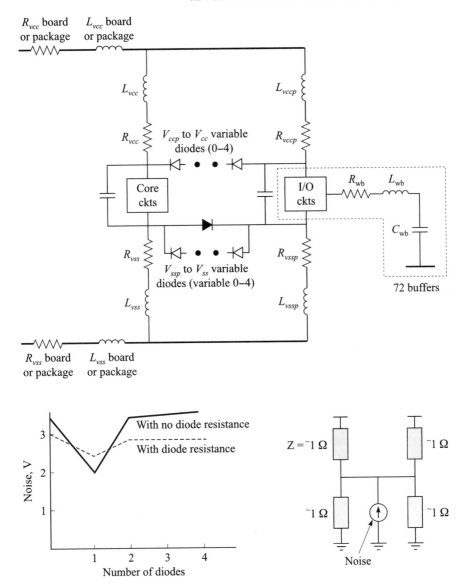

Figure 6-8. The ESD coupling diodes can couple some noise from the I/O to the core supply. In this example the core power supply has a high impedance; therefore, addition of coupling diodes effectively brings down the system impedance. This reduction in turn lowers the core power supply noise.

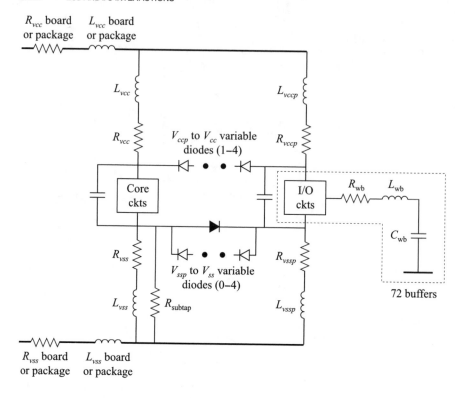

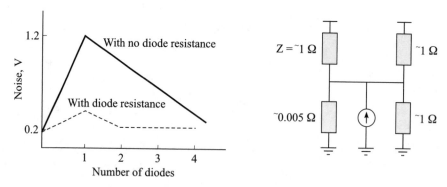

Figure 6-9. The ESD coupling diodes can couple some noise from the I/O to the core supply with substrate tap.

As noted earlier, the switching times for the I/O, core and di/dt components are different. It may be a good choice to merge the power supplies, instead of the two, to create one "time-shared" low-impedance system and remove the ESD protection for each supply.

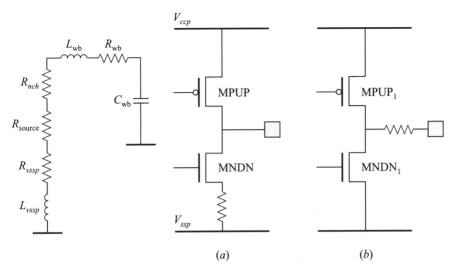

(a)   (b)

Figure 6-10. (*a*) By adding a source resistor, the *di/dt* noise in V_{ssp} is reduced and the noise is damped strongly [Senthinathan, 1994]. The source resistor implementation can help limit the switching current noise and damp the V_{ssp} noise; however, the source resistor will reduce the ESD hardness. (*b*) Placing the resistor on the drain side is an acceptable method to dampen the power supply transient but provides no negative feedback as the source implementation.

This section has discussed interactions between peripheral and core noise due to ESD diodes for the driver chip. There are also ESD- and I/O-related interactions at the receiver. These interactions need to be examined.

In a real I/O circuit, coupling of V_{ssp} to V_{ssc} may be more important than coupling of V_{ccp} to V_{cc} since there is only one diode (parasitic) between V_{ss} and V_{ssp}. However, for V_{cc} power supplies, it is possible to completely separate the diodes or string along a large number of coupling diodes so more noise immunity is possible.

In practical circuits, a source resistance on the NMOS reduces *di/dt* noise [Senthinathan, 1994]. The two contributing factors are the weak debiasing effect of the gate source at high currents and the damping of the *LRC* power supply ringing. A circuit schematic showing such a scheme is shown in Figure 6-10. However, this circuit may not be ESD robust, as discussed earlier in Section 5.1. Rather a more robust method is to introduce the resistor on the drain side.

6.3. ESD DIODES AND I/O SIGNAL INTEGRITY

The ESD diodes also impact the I/O signal integrity to a degree, and the result is usually beneficial. In this section the interaction of the ESD diodes and signal clamping is discussed. This is especially helpful on buses that have to be over-driven because they are heavily loaded and have to meet aggressive timings.

When a transmission line is driven with a source resistance having exactly the same impedance as Z_0, half the voltage step is coupled into the line. This voltage wave travels to the end of the line and reflects back with twice the voltage, making it equal to the full rail voltage. This reflection comes back to the driver and then restores the driver to the full voltage value.

Consider such a driver that is exactly matched at the slow condition. There are no under- or overshoots in such a system. However, in the fast condition the output resistance R_0 is now significantly lower (maybe $\frac{1}{2} Z_0$) and a signifi-cant overdrive occurs. This voltage is again doubled, and it arrives back at the driver with a magnitude 1.33 times the rail voltage, which is now not perfectly terminated. Thus two things happen, there is an overvoltage developed and there are multiple reflections. Typically, even in the slow condition the driver is set to higher than the exact $\frac{1}{2} Z_0$ value, say $Z_0/2.25$. Thus the reflections cause large over- and undervoltage as the fast condition. This is shown in Figure 6-11.

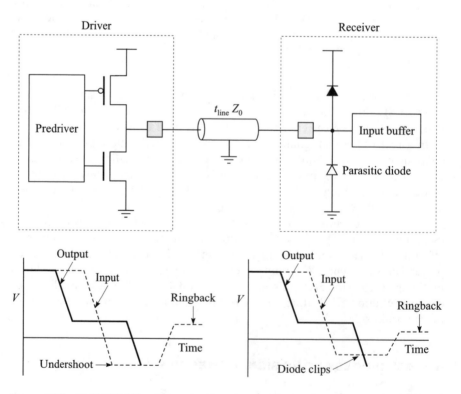

Figure 6-11. A CMOS I/O scheme showing the two cases with and without diode clip-ping.

The I/O schemes discussed above are reflected-wave switched schemes. They need at least twice the propagation delay to complete the switching all along the line, and they are good for slow systems. However, for faster I/O schemes, incident wave switching is used. In incident wave switching, the first edge causes the switching, and reflections are not necessary to cause the switching. These call for significant overdrive, maybe 75% of the full V_{ccp} step. As can be seen in Figure 6-11, the overdrive causes even larger over- and undervoltages due to reflections in a system.

In a system with a CMOS driver and a receiver, there will be at least one parasitic diode, as shown in Figure 6-11. Here the P-epi process is shown with a corresponding diode from the pad to V_{ss}. This diode provides clipping action for undershoots. If there is a diode present between the pad and V_{cc}, there will be clipping for the overshoots as well [Annaratone, 1986; Dabral, 1994].

This clipped wave reflects back once again at the driver. At the driver end, the wave is terminated by a lower impedance (compared to the transmission line). This time the voltage reverses and is reduced. This wave travels, and when it impinges on the last receiver in the line (an open circuit) it doubles and causes a ringback. If there are any transistor–transistor logic (TTL) inputs in the system, then the reflection from the driver ringback should not exceed 0.8 V, which is the V_{il} for TTL. However, the diode clamping in the first reflection at the receiver helps in reducing the ringback effect, as can be seen in Figure 6-12.

This clipping action can inject a significant current into the receiver power supply, which causes significant noise in the core supply. To understand the ringback phenomenon, consider the circuit in Figure 6-13. It consists of a low-

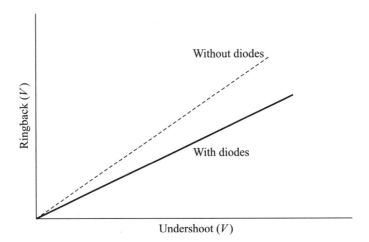

Figure 6-12. Ringback as function of undershoot. The undershoot is the (negative) voltage at the receiver without the diode, and the ringback is the (positive) voltage at the receiver.

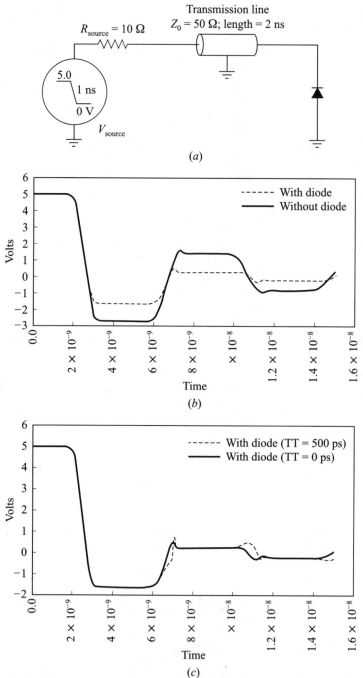

Figure 6-13. (*a*) Simulation setup to measure ringback. (*b*) Ringback is reduced when a diode clamp is utilized. (*c*) Transit time affects amount of charge stored, which then needs to be cleared before diode can be forward biased.

impedance signal generator driving into a typical transmission line and terminated by a diode. This is a very simple model of a driver–receiver pair.

Now the undershoot and ringback can be correlated using simple transmission line theory. The undershoot without a diode is given by the equation

$$V_{us} = V_{dc}\left(\frac{R - Z}{R + Z}\right) = V_{dc}\rho, \qquad (6\text{-}4)$$

where V_{us} is the undershoot when no diode is present, V_{dc} is the voltage of the I/O system, R and Z are the source resistance and impedance of the transmission line, and ρ is the resulting reflection coefficient at the driver [Dabral, 1994]. The positive ringback voltage, shown in Figure 6-13b, is

$$V_{rb} = V_{dc}\rho^2 \qquad (6\text{-}5)$$

When the diode is present, it acts as a dynamic termination. Assuming it begins clipping at 0.7 V and has effective resistance R_{diode}, the reflection coefficient Γ is given as

$$\Gamma = \frac{V_{out}}{V_{in}} = \frac{R_{diode} - Z}{R_{diode} + Z} - \frac{(0.7)Z}{(R_{diode} + Z)V_{in}} = \frac{R_{diode} - Z}{R_{diode} + Z} - \frac{2(0.7)Z}{(R_{diode} + Z)V_{dc}\rho} \qquad (6\text{-}6)$$

The reflection coefficient Γ is the ratio of outgoing to incoming traveling waves at the receiver. If there is no diode resistance, note the resulting simple form that gives the required $V_{out} + V_{in} = -0.7$ V at the diode. Coefficient Γ is derived by having the diode current $I = (V_{out} + V_{in})/Z$ and diode voltage $V = V_{out} + V_{in}$ satisfy the resistive ideal diode I–V relation. If there is no diode (R_{diode} infinite), $\Gamma = 1$ and the entire current is reflected back to the driver. The last part of Eq. 6-6 puts in the receiver's initial incoming wave $V_{dc}\rho/2$ to obtain the wave reflected back to the driver. After another reflection ρ from the driver, the ringback voltage V_{rb} is then given by

$V_{rb} = $ (initial wave)(Γ)(Reflection at driver)(1 + reflection at receiver)

$$= \frac{V_{dc}\rho}{2}\Gamma\rho(2) \qquad (6\text{-}7)$$

$$V_{rb} = \frac{(R_{diode} - Z)V_{dc}\rho^2}{R_{diode} + Z} - \frac{2(0.7)Z\rho}{(R_{diode} + Z)} \qquad (6\text{-}8)$$

There is good agreement between simulated and calculated ringback (Figure 6-13) because R_{diode} is high enough that Γ is positive, and the fact that charge storage in forward bias is of minor consequence here [Dabral, 1993]. For high undershoots the diode is heavily forward biased and charge storage effects must be considered.

When heavy forward bias results from low R_{diode} and severe mismatch between line and driver, the diode resembles a step-recovery diode [Sze, 1969] although the phenomenon of interest is not the speed of the voltage rise when the diode comes out of forward bias but the size of that positive voltage, due to the reflected waves being amplified as the stored charge is pumped out of the diode. A detailed treatment of ringback as caused by heavy forward bias, charge storage in the diode, and pumping out of that charge [Maloney, 1993] is beyond the scope of this work, but with the general principles discussed above concerning traveling waves up and down the line, it is possible to develop some idealized mathematical expressions that capture the phenomena pretty well. The concepts can also be put into a spreadsheet program that develops and follows the wave series one time step ($2L/c$, $L =$ line length) at a time, resulting in ringback. While some of the forward bias current is known to result in stored charge, exactly how much does, and what a decay time for it might be (all relating to transit time as discussed above), is not known *ab initio* and must be fit to time-dependent experimental data. The spreadsheet program is particularly useful for finding a credible fit for these parameters.

Here is a simple example of a wave series. Using quantities as defined earlier, suppose, for mathematical simplicity

$$Z = 40\,\Omega \quad V_{dc} = 5\,\text{V (to be switched to 0 V)}$$
$$R = 10\,\Omega \quad V_{diode} = -0.5\,\text{V}$$

Then $\rho = 0.6$ at the driver, and for an ideal diode $\Gamma = -1 -0.5/V_{in}$, whenever $V_{in} < -0.25\text{V}$; V_{in} is the magnitude of the incident wave. $\Gamma = +1$ otherwise, unless charge is being pumped out.

These reflection coefficients allow boundary conditions at the ends of the line to be satisfied. In terms of incoming and outgoing waves,

$$V = V_{in} + V_{out} \qquad I = (V_{in} - V_{out})/Z \qquad \rho \text{ and } \Gamma = V_{out}/V_{in}$$

Boundary conditions are the following: At the driver, $V/I = R$. Diode voltage is $-0.5\,\text{V}$ or greater, $R_{diode} = 0$.

The initial steady 5 V is decomposed into 2.5 V waves traveling in opposite directions and reflecting from terminations with $\rho = 1$. Then the wave series is as in Figure 6-14.

Notice how incident and reflected waves at the driver are always in ratio -0.6 and how waves at the diode are reflected to maintain the diode drop of $-0.5\,\text{V}$. This will continue even after the wave has been cut down to below half a forward bias if there is a charge storage, but notice that the current (proportional to $V_{in} - V_{out}$) then reverses, pumping the charge out by amplifying the waves until the charge is gone. Only then does $\Gamma = 1$ for the diode, but by then the wave is positive and the voltage doubles. This is the ringback pulse, delayed

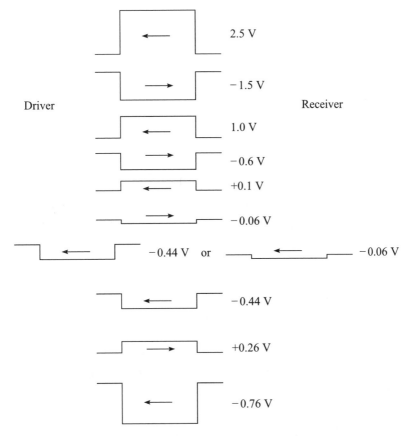

Figure 6-14. Wave series for driver and diode receiver. If charge is stored during the initial forward bias phase, the diode drop −0.5 V continues to be maintained as charge is pumped out.

several time steps, which can cause logic upset. The diode voltage suddenly changes from forward bias to a substantial reverse bias, resembling the pattern shown in Figure 6-15.

Finally, if the I/O signaling voltage and V_{cc} are not the same, then the symmetry for the low-going ringback and clamping is not similar to the high-going transitions. This occurs because V_{ss} and V_{ssp} are the same, only split, whereas V_{ccp} and V_{cc} are different. Therefore the V_{ss} diode will turn on any time the pad voltage undershoots the diode turn on voltage. However, if the V_{cc} is sufficiently high, the rising edge will not see any clamping. This leads to an asymmetric "effective" net response that may cause data jitters, reducing the valid data window. Such unbalanced cases are seen in an open drain, or low-voltage I/O and on a high-voltage core.

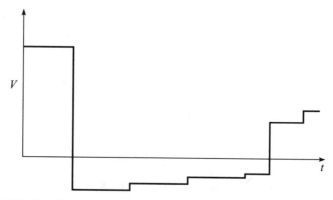

Figure 6-15. Diode voltage versus time in a case causing heavy forward bias, charge storage, and a reflection amplifier effect. The charge is finished being pumped out in the middle of a time step.

6.3.1. High-Frequency Leakage in I/O

The ESD structures can interact with the I/O in a number of unexpected ways. Consider the grounded-gate device used in protecting a high-frequency clock input pad, as shown in Figure 6-16 [Weston, 1992]. The grounded gate had enough β and coupling through the capacitance to forward bias the NPN transistor. This caused significant leakage at the input pin. Since this *RC* circuit is a high-pass circuit, with increasing frequency more coupling should be expected.

The methods to suppress the leakage are to decrease the substrate voltage below zero. This effectively debiases the BJT and leakage is reduced. However, in a CMOS technology this option is not possible because the P-epi is tied to ground. It can also be argued that in CMOS technologies of the future the

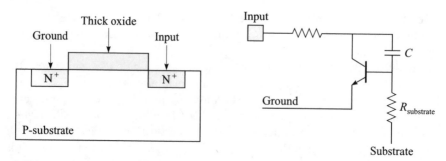

Figure 6-16. An NPN parasitic transistor may leak current for high-frequency input. The high substrate resistance and sufficient coupling through the capacitance enables the BJT to turn on.

substrate resistance will be sufficiently low that the BJT can never be forward biased. The other option is to increase the N^+-to-N^+ spacing in the TFO, thereby decreasing β. However, this will also change the trigger voltage of the clamp and that needs to be factored in.

This example has parasitics similar to the GCNMOS scheme. The pole of the high-pass filter for GCNMOS is at a much lower frequency when compared to this case. Thus it will couple high-frequency components earlier, in turn implying that for high-frequency I/O some modifications to the GCNMOS scheme will be required. These will effectively have to shift the pole to a much higher frequency during circuit operation.

6.4. SUMMARY

Electrostatic discharge and I/O interact at many levels. Usually ESD reliability imposes additional constraints in design and layout of the I/O buffers:

- The NMOS output sizing may be constrained by some ESD rules that may enforce granular width and fixed channel lengths, thereby limiting the I/O options in drive strengths.
- The granularity can affect the finest compensation settings possible. In addition, some schemes (e.g. N-well resistor for NMOS ballasting) make the point of compensation generation different than the compensation point of use.
- In terms of a pass transistor for high-voltage input-tolerant buffers, the best performance in speed leads to poorest ESD performance and vice versa. Thus a reasonable pull-up sizing for the pass transistor is important for optimum ESD and I/O performance.
- The periphery and core switching generate $L\,di/dt$ noise. Previously, the periphery was the noise supply compared to the core power supply. However, that may not be the case today and each chip should examine the periphery and core noise. If the peripheral nose is high, then there is a possibility that the I/O will couple some noise through the ESD diodes into the core.
- As the I/O voltage scales down and more compensation of I/O buffers becomes prevalent, the chances of the ESD diode coupling noise into the power supply decreases. The reduced core voltages also require lower noise, thus making diode turn-on extremely unlikely.
- The ESD diodes form good clamps to limit severe under- and overshoot in overdriven systems. These can inject significant current into the power supply of the receiver in the case of severely overdriven and multiload bus.

REFERENCES

[Dabral, 1993] S. Dabral, R. Aslett, and T. Maloney, "Designing On-Chip Power Supply Coupling Diodes for ESD Protection and Noise Immunity," *Proc. EOS/ESD Symp.*, 1993, p. 239.

[Dabral, 1994] S. Dabral, R. Aslett, and T. Maloney, "Designing On-Chip Power Supply Coupling Diodes for ESD Protection and Noise Immunity," *J. Electrost.*, **33**(3), 1994, p. 357.

[Gabara, 1996] T. Gabara, "A Closed-Form Solution to the Damped RLC Circuit with Applications to CMOS Ground Bounce Estimation," *IEEE ASIC96*, Sept., 1996, p. 73.

[Gabara, 1997] T. J. Gabara, W. C. Fischer, J. Harrington, and W. W. Troutman, "Forming Damped LRC Parasitic Circuits in Simultaneously Switched CMOS Output Buffers," *IEEE JSSC*, **32**(3), 1997, p. 407.

[Hanafi, 1992] H. I. Hanafi, R. H. Dennard, C. L. Chen, R. J. Weiss, and D. S. Zicherman, "Design and Characterization of a CMOS Off-Chip Driver with Reduced Power-Supply Disturbance," *IEEE JSSC*, **27**(5), 1992, p. 783.

[Maloney, 1993] T. J. Maloney, "Ringback Summary," Intel internal memo, Feb. 1993, unpublished.

[Senthinathan, 1994] R. Senthinathan and J. L. Prince, *Simultaneous Switching Noise of CMOS Devices and Systems*, Kluwer Academic, Boston, 1994, p. 96.

[Sze, 1969] S. H. Sze, *Physics of Semiconductor Devices*, 1st ed., Wiley, New York, 1969, p. 137.

[Tong, 1996] M. Tong, R. Gauthier, and V. Gross, "Study of Gated PNP as an ESD protection Device for Mixed-Voltage and Hot-Pluggable Circuit Applications," *Proc. EOS/ESD Symp.*, 1996, p. 280.

[Weston, 1992] H. T. Weston, V. W. Lee, and T. D. Stanik, "A Newly Observed High Frequency Effect on the ESD Protection in a Giga Hertz NMOS Technology," *EOS/ESD Symp.*, 1992, p. 95.

[Wong, 1988] D. T. Wong, R. D. Adams, A. Bhattacharya, J. Covino, J. A. Gabric, and G. M. Lattimore, "An 11ns 8K×18 CMOS Static RAM with 0.5mm Devices," *IEEE J. Solid State Cir.*, **23**(5), 1988, p. 1095.

CHAPTER 7

MIXED-VOLTAGE ESD

Chapter 6 discussed primarily the power supply, ESD, and I/O interactions in a single-voltage system. Power delivery may have been split for noise isolation consideration but voltage levels were the same. When different voltage levels exist between the I/O and the core supplies, a special set of issues develops. These will be examined in this chapter.

In the past, 5-V TTL CMOSs were primarily used in system design. Even when different technology generations were used in system subcomponents, voltage levels were the same (5 V). With aggressive device scaling, voltages have had to decrease correspondingly. Some early standard 5-V systems such as CMOS DRAMS changed from 5 to 3.3 V [Prince, 1992]. Thus, system voltages were no longer consistently 5 V but mixed 5 and 3.3 V, and special design considerations were required.

New I/O scenarios were subsequently encountered. For example, the memory system could be 3.3 V and the processor 5 V. Voltages have scaled down consistently, and today 3.3 V is common, 2.5 V is in design, and lower than 1.3 V is being discussed. This trend in lower voltages is expected to continue, resulting in some systems having mixed-voltage I/O. The mixing of voltages has an impact on ESD protection, and therefore it must be understood. The basic scenarios of mixed voltage are shown in Figure 7-1. The main issues that result can be grouped into four main categories:

- input reliability, timing, and leakage;
- output reliability and predriver;

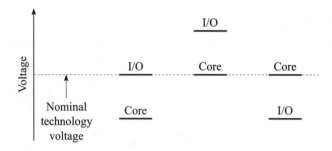

Applications	Mobile low power	Backward compatibility	Low power I/O and fast I/O
Issues	Voltage translators required at output predriver	Voltage translators required at output predriver Input clamp required Output buffers High-voltage compatible	Input voltage translation

Figure 7-1. Mixed-voltage scenarios and their motivations and major issues.

- power distribution protection on the chip; and
- system level issues.

In this chapter, some techniques to address these issues will be discussed. The emphasis will be on designable protection.

7.1. MIXED-VOLTAGE INPUT DESIGN

Input designs are altered significantly in a mixed-voltage scenario. There are two cases, one where the core voltage is higher than the I/O and the other where the core voltage is lower than the I/O voltage. In the first case, the low I/O voltage cannot turn off the PMOS completely in the input buffer, and special considerations have to be made. In the high-voltage I/O input buffer reliability is threatened and must be addressed. A whole host of issues then result.

7.1.1. Input Design for Low-Voltage Core and High-Voltage I/O

In the current path methodology, a dual diode strategy has been discussed as a designable approach to ESD protection. The dual-diode strategy works well when the I/O and core voltage are at the same voltage level. However, when the on-chip voltages are lower than the I/O voltage, a simple dual-diode strategy will forward bias the pad to the core V_{cc} diode and will clamp the I/O voltage. Instead of a single diode between the pad and V_{cc}, a diode string has to be employed. This is shown in Figure 7-2.

Stringing diodes has an impact on area utilization and resistance of the ESD path. The diode is in series (n in number) so the effective diode resistance increases by n. To reduce resistance down to a single-diode value, the area of each diode has to be scaled up by a factor of n. Therefore the total area of the diode chain is then n^2 times (n diodes having n times the area) the area of a single-diode implementation [Voldman, 1994]. This quick analysis assumes that the diode current voltage is dominated by the resistive component. If the diodes act as PNP transistors, then the sizes do not have to scale as n^2 but rather to a smaller degree, as estimated by Eq. 2-23. Also as the general technology voltage decreases with time, the differences in the mixed-voltage interfaces decreases. Concurrently, the diode cut-in voltage remains the same, thereby greatly reducing the number of diodes to be strung in series. For example, if the cut-in voltage of a diode is 0.4 V, then for a 5 V I/O reference to a 3.3-V core would require at least four diodes, whereas for a 3.3-V I/O referenced to a 2.5-V core would require only two diodes.

Another important addition necessary to protect the gate oxide of the input buffer from the high-voltage I/O is to place an NMOS device in the input path. This pass transistor MN_1 has its gate tied to V_{cc} and so limits the voltage on input buffer node N_1 to a maximum of $V_{cc} - V_{tn}$. Beyond this voltage at N_1, the NMOS MN_1 enters the subthreshold region and effectively starts turning off.

Since node N_1 never reaches the V_{cc} level, the inverter IN_1 is never completely off when input is driven high as the PMOS will be in the subthreshold region. To turn the inverter completely off, the PMOS must be cascaded with

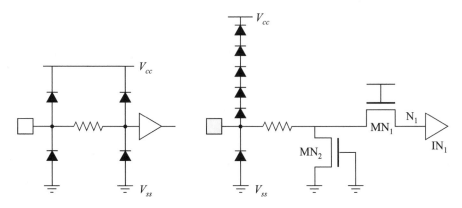

Figure 7-2. Additional complications and area penalty in supporting high-voltage I/O in low-voltage technology.

an enable signal or other pull-up techniques. The consequences can lead to larger and slower input receivers.

The pass transistor affects I/O performance in two ways: by increasing absolute delay through the path and by increasing the signal jitter and quality. Usually jitter reduction is the prime concern for clock circuits and delay reduction is the prime concern for data input circuits. First clock circuits are discussed followed by data input circuits.

Consider the feedback path in a phase-locked loop (PLL) or a delay-locked loop (DLL) scheme [Waizman, 1995], as shown in Figure 7-3. In this scheme an on-chip-generated signal is compared to an external reference clock for the purpose of aligning the chip clocks to a system clock [Gardner, 1980]. If the feedback path is directly compared to the incoming reference clock signal, as shown in Figure 7-3a, then the clock tree is skewed from the reference clock by an input delay. This input delay is δ_1 for chip 1 and δ_2 for chip 2. The input delay depends on the process (i.e., fast or slow part), voltage, and temperature of the two chips. Therefore, the two clock trees on the two chips experience a skew that is the difference between the two input delays. To reduce this skew, the feedback path is also made to go through a similar input path, as shown in Figure 7-3b. This delays the feedback path by an equivalent delay experienced by the reference clock, thereby allowing both clock trees in the two chips to align well to the reference clock. This drastically reduces the systematic skew, thereby allowing better synchronization between the two clocks. One additional important criterion is the availability of sharp external and internal clock edges to allow a high-resolution comparison, which in turn aligns the phases of the two clocks.

In Figure 7-3c, the feedback path for a PLL reference and generated clocks is shown. This case uses a high-voltage clock driver whereas the receiver is a lower voltage chip. This high-to-low voltage conversion needs a pass transistor MN_2. However, with MN_2, the signal paths and the voltage levels for the feedback and input signals are different and will cause additional clock skew between the feedback and reference clocks. This can be seen in the voltage levels and timing.

The skew may be reduced by matching the paths better. A pass transistor can be added to the feedback path, which will allow better matching. However, some skew will still be expected because (a) the input reference clock is driven by a high voltage (V_{cch}) and the internal clock is limited to V_{cc} and (b) there are mismatches in the intrinsic pass transistor. If the input paths are not well matched, the signal characteristics are different (i.e., rise and fall times, duty cycle) and lead to increased skews. As seen in Chapter 4, any clock unpredictability, such as skew and jitter reduces the allowable communication window or increase the latency. This scheme will reduce the systematic skews to a large extent; however, the much smaller signal jitters have to be accepted. With high-speed I/O even small tens of picoseconds are important, and this trend will continue.

For the PLL, there is feedback control, so the small skews and jitters in the input path may only cause small errors, primarily because there is a feedback path that compensates for the pass transistor distortions. A more significant penalty can result when the pass transistor is in a normal data input path. This can be seen in the timing diagram in Figure 7-4, where the duty cycle is altered. The time to propagate a high-to-low input signal is less than the low-to-high-going input signal. For example, consider an input signal at 50 MHz (i.e., 20-ns period, 10 ns low and 10 ns high) with the I/O at 5.5 V (5 V + 10%), the core at 3.0 V (3.3–10%), an input ramp rate of 2 V ns, a $V_{tn} = V_{tp} = 0.8$ V. The trip point is set to 2.75 V (half of 5.5 V). If the receiver is at 3.0 V, the voltage level reached by the receiver inverter is 2.2 V, clearly not sufficient to meet the intended trip point of 2.75 V. An alternative input buffer trip point at 1.1 V (i.e., half the voltage swing to maximize the noise margin) may be acceptable. The input low-to-high transition is then delayed by $(5.5\,\text{V}-1.1\,\text{V})/(2\,\text{V/ns}) = 2.2$ ns and the high-to-low transition by $1.1\,\text{V}/(2\,\text{V/ns}) = 0.55$ ns. Therefore the duty cycle is skewed from 10 ns high and 10 ns low to 11.65 ns high and 8.35 ns low. This also affects the common clock operation, as the delays to propagate a high and a low are skewed by 1.65 ns.

It can be argued that it is possible to skew the driver such that the receivers see a 50% duty cycle at a 1.1-V trip point. However, if there are 5-V circuits with an existing 2.5-V trip point, they will suffer duty cycle variations. So the problem point just moves to the higher voltage system.

The timing implications on input design of the mixed-voltage schemes have been discussed. There are additional DC concerns in the scheme, such as leakage currents in the input buffer and oxide reliability concerns [Hu, 1994]. The leakage current through the MN_1 transistor can cause high-voltage buildup at the inverter input. This leakage is caused by subthreshold conduction that is not sufficient for quick charge-up but given sufficient time will cause the input node voltage to increase towards the external high-voltage level. The high voltages can then stress and degrade the input buffer's gate oxide and therefore affect the performance. Another timing-related factor is that since there is an unknown amount of charge-up at the input gate, the delay during discharging will vary according to the excess charge. The discharging of the excess charge (above $V_{cc} - V_{tn}$) will cause extra delay, which will lead to jittering in the inverter output by small amounts, depending on the time duration of charge-up and the final level of the input gate charge-up. This skew is shown in Figure 7-5.

A solution to this leakage charge-up is to provide clamping at the input node. One such scheme is shown in Figure 7-5. A weak P pull-up device MP_1 can ensure that the node N_1 never rises above V_{cc} and that it is also pulled up to V_{cc} [Greenhill, 1997]. This also ensures that the input buffer PMOS is completely turned off and the inverter gate is protected. The jitter caused by the excess charge-up is thus avoided. The result is improved timing repeatability and oxide reliability. Similar pull-up schemes to reduce delay can be found in TTL CMOS inverters [Yoo, 1995].

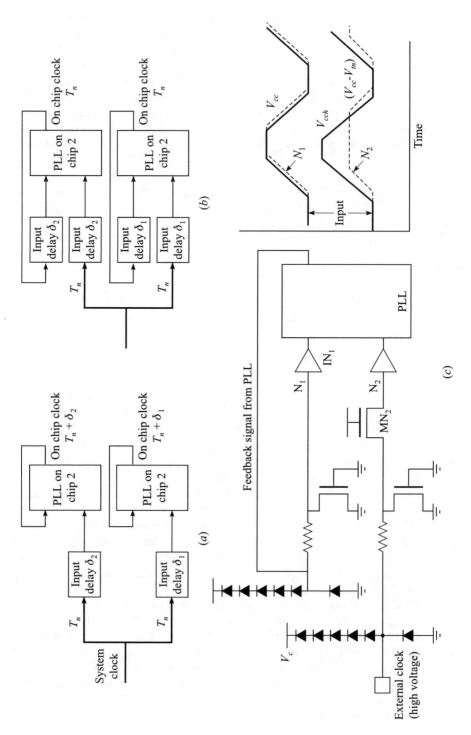

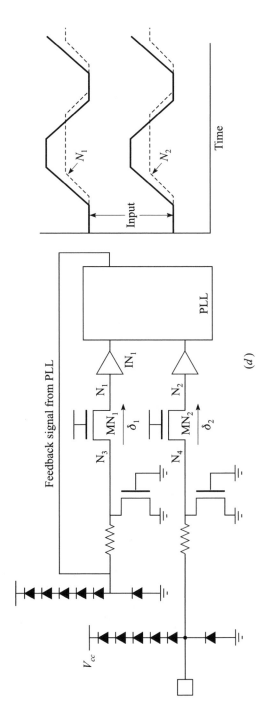

Figure 7-3. A DLL/PLL skew and jitter reduction scheme: (*a*) when the feedback path is not matched to the input path, maximum skews between two chip clocks occur. (*b*) The skew can be reduced by introducing the exact delay as in the reference clock path. (*c*) The feedback path mismatch in a high-voltage reference clock and a low-voltage feedback clock. (*d*) There is skew reduction in the paths by matching the inputs better, thereby reducing the systematic skew but increasing the jitter component [Young, 1992].

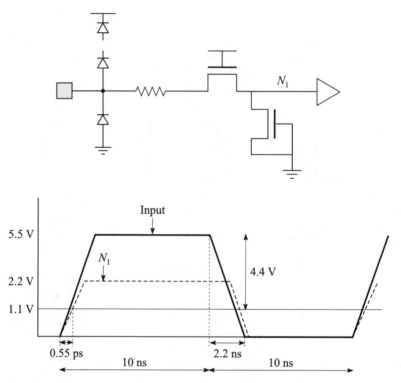

Figure 7-4. Signal skew due to high-voltage interface. Input buffer trip point is set to 1.1 V to maximize noise margin.

It is the subthreshold conduction through the pass transistor MN_2 that causes the leakage that leads to the charge-up of the input node N_1. The subthreshold conduction can also be used to remedy this issue. A scheme taking advantage of this concept is shown in Figure 7-6. By adding a transistor MNCL in parallel to the inverter input, another subthreshold path to ground is established [Chen, 1996]. The voltage at node N_1 stabilizes when all the leakage current coming in through the NMOS MN_1 exits from the NMOS $MNCL_1$. When there is no leakage path established to ground, the circuit behaves as an *RC* (nonlinear, high-resistance) circuit and charges up slowly to near the externally applied voltage. However, when the serial path to ground is established, a voltage division action occurs (between two high-resistance paths created by MNPT and MNCL), and the voltage is clamped to an intermediate value where the leakage charging current going in through MNPT is equal to the discharging current out to MNCL.

The functioning of this scheme can be seen in Figures 7-6 and 7-7. As the voltage at node N_1 increases, the V_{ds} and V_{gs} across the pass transistor MNPT decreases, reducing the subthreshold leakage. If the voltage at N_1 exceeds the V_{cc} level, then for MNPT, V_{gs} is negative. The subthreshold leakage is an

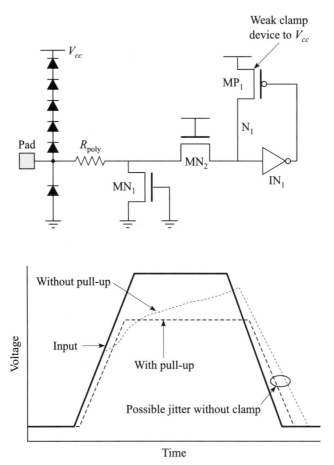

Figure 7-5. Adaptation for 5-V tolerant input structure ensuring no high voltage at input nodes. The clamping PMOS MP_1 clamps N_1 to V_{cc} and ensures a known starting point, thus resuming jitter.

exponential function of V_{ds} so the decrease in leakage is large. Simultaneously the V_{ds} across the transistor MNCL increases, which increases the leakage conduction. The V_{gs} across MNCL is zero. Thus depending on transistor sizing and the process conditions, a stable clamp level will be reached when the decreasing current of MNPT matches the increasing current of MNCL.

Theoretically this subthreshold current division is fine; however, when noise charge is injected into N_1 due to signal coupling noise or power supply ripples, it will persist there for considerable time because subthreshold discharging becomes extremely small. Another issue with such a scheme is that if the pass transistor is damaged (ESD, manufacturing, EOS), it can allow conduction far more than based on the subthreshold criterion [Zupac, 1992], and it can easily overpower the discharging (MNCL) transistor. In view of such

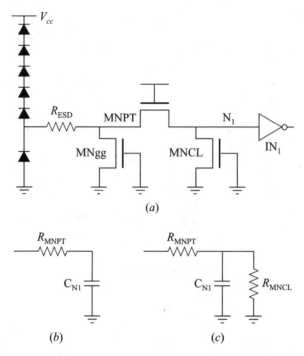

(a)

(b)   (c)

Figure 7-6. Subthreshold technique to ensure that no node is above the oxide stress voltage: (a) when no NMOS clamp MNCL is used and (b) when MNCL is placed.

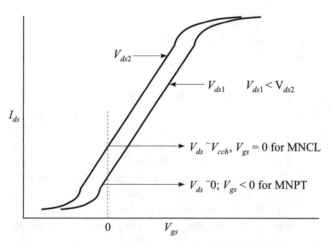

Figure 7-7. Higher body biasing of pass transistor MNPT makes it less leaky than the clamp MNCL. ESD damage to MNPT or noise in V_{cc} supply can leak more current than expected.

uncertainty and the comparative ease of design of the weak pull-up scheme, weak pull-up should be preferred.

Additional methods exist to handle high-voltage inputs. One simple method is to use a resistive divider to sample a fraction of the input voltage that is then fed to the input receiver (this is fairly common in high-voltage measurement and in oscilloscope sampling). This approach is shown in Figure 7-8. The implementation of these resistors on-chip involves trade-off between area, capacitance, and allowable leakage current.

This section has discussed the situation when the high-voltage input is applied to a low-voltage core. The next section will examine the situation where the low-voltage signal is applied to a high-voltage core.

7.1.2. Input Design for High-Voltage Core and Low-Voltage I/O

When the I/O voltage is lower than the core voltage (lower by V_{tp}), the PMOS in the input buffer cannot turn off fully. This is shown in Figure 7-9, where the PMOS MP_1 leaks when the input is high. When the input is low, the NMOS is off and no leakage occurs. This causes leakage current and asymmetry in the input transfer characteristics. This issue was also encountered in the pass transistor clamp inputs discussed in Section 7.1.1. Here a PMOS pull-up device was used to bootstrap the buffer input to V_{cc} level so that the voltage buildup due to NMOS leakage was curtailed. However, doing so in this case will result in a direct leakage path between the driver and the pull-up device. A solution is to use differential receivers. Here the input voltage is compared to a reference voltage (V_{ref}) and the output toggles accordingly. The differential input also dissipates DC power, but it has significantly better transfer characteristics.

For a complete CMOS operation (no leakage) a voltage translator circuit is required. These are discussed later in Section 7.3.

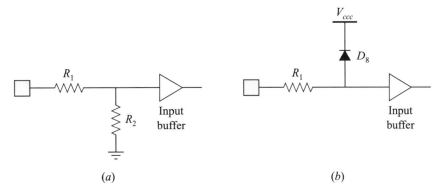

(a)   (b)

Figure 7-8. High input voltage is reduced by (a) a resistive divider and (b) a resistor diode combination [Zanders, 1991]. There is a DC path in both cases, leading to some wasted power. By suitably increasing the resistor value, the leakage can be minimized but the capacitance and area need to be increased.

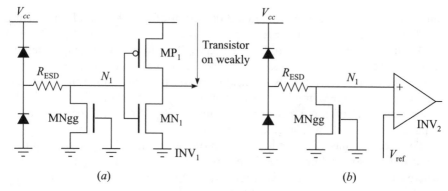

Figure 7-9. (*a*) Issue with a low-voltage input causing leakage and (*b* a differential input where a reference voltage (typically half of the voltage swing) is compared to the I/O voltage.

7.1.3. Capacitive-Coupled Input Receiver

The effect of either low- or high-voltage input needs to be compensated for in the design of the circuit. However, if the input node was only AC coupled into the chip and the DC was blocked out, then the voltage level sensitivities would be of little consequence. This functionality can indeed be obtained using a capacitance DC blocking element [Gabara, 1997].

Consider the circuit shown in Figure 7-10, which shows the input node coupled into the chip using a capacitor configured as a high-pass filter. This circuit passes the transients into the chip but blocks the DC. The received voltage can be used to retrieve the input. However, two issues exist before such a scheme can be successfully implemented: the physical area of the capacitor and the DC level decay with time across the input resistor. The capacitor can be formed under a pad using the polysilicon and metal. Thus the DC level decay issue needs greater consideration in the circuit design.

The decay of the input level will ultimately cause any input receiver to be in a wrong state. A method to address this issue is to use a resistor that creates an *RC* time constant shorter than the smallest pulse width. This scheme will avoid any input data pattern dependency. To create a steady received signal, a com-

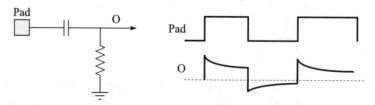

Figure 7-10. Capacitor blocking scheme indicating decoupling of DC levels in system, also shown is DC "level wandering."

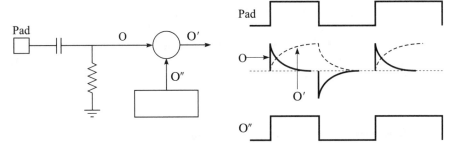

Figure 7-11. By addition of complementary exponential wave to decaying input pulse, a steady DC input pulse can be created [Gabara, 1997].

plementary exponential pulse can be added to the received signal, as indicated in Figure 7-11 and termed quantized feedback [Gabara, 1997]. This input circuit has been effectively utilized in interfacing GTL, PECL (pseudo-ECL), HSTL, and LVDS to a single core receiver design.

Having examined the input buffer options, the effect of mixed voltage on output drivers and predrivers is examined next.

7.2. OUTPUT DESIGN FOR MIXED VOLTAGE

The output driver designs can be categorized into two styles: high-voltage drivers (which actively drive higher voltage than the core voltage) and high-voltage-tolerant drivers, which drive only low-voltage signals but must be tolerant to high voltage when they are in the receive mode. This is illustrated in Figure 7-12. In this section the effects of mixed-voltage factors on the open-drain and CMOS buffer styles will be discussed.

In both cases the single devices in the output driver that were simple N or PMOS have now to be cascaded in the high-voltage-tolerant or drive cases

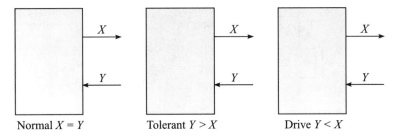

Figure 7-12. Difference between tolerant and drive I/O. In tolerant I/O the receive is high voltage (typically higher than the technology voltage) but the drive is lower. In driver case both the receive and drive may be at the high voltage.

[Krakauer, 1992; Greenhill, 1997]. This cascading has to be done such that the gate–drain voltage never exceeds the prescribed safe levels, as shown in Figure 7-13. This works well for a high-voltage drive and receive scheme. The direct impacts of cascading are increases in driver size (i.e., a factor of approximately 2), longer delays, larger predrivers, and increased layout complexity.

7.2.1. Open-Drain Design for High-Voltage I/O

It was previously assumed that when a high I/O voltage was used, a high-voltage driver was also required. For ESD protection a single diode could be placed between the pad and V_{ccp}. What happens when the positive I/O supply is smaller than the external pad voltage or when it is not present (as in the case of open drain systems)? An example is shown in Figure 7-14. If the high V_{ccp} is not available, then in the diode protection method the single diode has to be replaced by a diode chain to prevent forward biasing the pad-to-V_{cc} diode. This chain can be designed and optimized as described in Chapter 2. Also, the chain can be tapered to minimize the area and capacitance and improve ESD functionality.

The other important factor to note is the addition of a clamp device to the common node in the cascade to limit the voltage buildup at node N_2. When the input IN is low, the NMOS MN_2 is off and the pad is pulled high by the external pull-up resistor. The top NMOS MN_1 is still on and charges N_2 to the $V_{cc} - V_{tn}$ level, at which point it starts shutting off. There is leakage through the top transistor charging up node N_2. In case the NMOS device

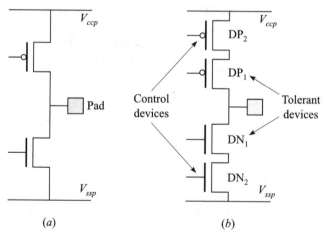

(a)   (b)

Figure 7-13. Typical comparison of low-voltage CMOS driver to high-voltage supporting driver. Biasing of nodes DN_1 and DP_1 has to comprehend oxide stress as well as device drive strength and size.

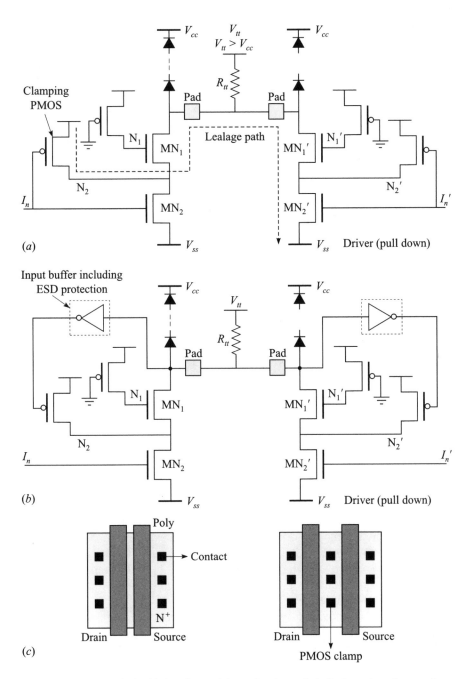

Figure 7-14. Open-drain high-voltage driver circuit noting diode string that replaces single diode. Also identified is a low leakage current path when implemented with (*a*) a simple pull-up for node N_2 and (*b*) a simple fix using an inverter. (*c*) The effect on the NMOS device layout is shown where the spacing between the poly increases to accommodate contacts for the PMOS clamp.

MN_1 is slightly damaged by EOS or ESD, the N_2 voltage can rise rapidly to the high I/O voltage.

In the cascaded driver, a natural subthreshold voltage division action is present due to the lower transistor. The discussion of subthreshold currents in Section 7.1 is valid here. If both transistors are undamaged, then the subthreshold voltage division prevents the N_2 voltage from building up to damaging levels. However, it is possible that the top transistor may be weakly damaged due to ESD (more likely than the input), in which case, it will leak more than the lower transistor capable of discharging, in which case the N_2 voltage may build up dangerously. For this eventuality the PMOS clamping transistor is required. Using this technique an open-drain device can be made to have sufficient ESD and oxide reliability.

A point to consider when making the leaker fix for node N_2 is that if it is not done right it can cause a leakage path. This is illustrated in Figure 7-14 for the case where two chips are connected and one pulls down the bus line. There is a leakage path established from one chip to the other. This path is between the pull up PMOS on the receiver, MN_1', and MN_2' on the driver. The leakage is usually small, but it can be bothersome in testing. Normally, when in the receive mode, there should be very little leakage at all. One method to fix the leakage is to tristate the receiver PMOS clamp using the input node as shown in b. Now if the bus is pulled low, the PMOS is turned off and no leakage results.

7.2.2. CMOS Buffer Design for High-Voltage I/O

In the open-drain case, the driver had to provide a pull-down action but the pull-up action was done by an off-chip pull-up resistor. However, in a CMOS buffer two types of operations are encountered, one where the driver has to drive a high voltage and the other where the driver has to tolerate a high voltage but drives only lower voltages (typically V_{cc}).

Now consider the case of a CMOS buffer that has to tolerate high input voltage but may not have to drive high voltages. An example is a 5-V input and a 3.3-V drive. In Figure 7-15, a simple cascading scheme for both NMOS and PMOS devices is shown. This may take care of the oxide reliability issue. However, heavy leakage will occur due to the single diode inherent in the biasing of the N-well device.

A solution is to provide a switch that disconnects the N-well when the pad exceeds V_{cc} [Voldman, 1994]. This will prevent the diode from forward biasing and causing leakage. This self-biased switch can be constructed using a PMOS device that is used to bias the N-well and is shown in Figure 7-16.

The operation of the scheme is as follows. When the pad is high (i.e., greater than $V_{ccl} - V_{tp}$), the PMOS MPNW turns off, disconnecting the N-well from the V_{cc} supply. As the pad voltage increases, the N-well charges up by the parasitic diode D_1 above the V_{ccl} level. For this reason the gate voltage of the two pull-up PMOSs should be kept at V_{ccl} to avoid stressing the gate

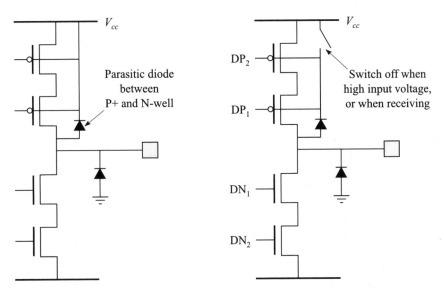

Figure 7-15. Parasitic diode in PMOS causes leakage paths when high voltage exists on the pad. So a switch must be used to disconnect the N-well from the V_{cc} supply to avoid leakage.

oxide. When the pad voltage decreases, MPNW turns on. This allows the N-well to discharge to the V_{cc} level. An additional circuit must be added to ensure that the N-well then discharges to the V_{ccl} level before the I/O buffer changes over from an input device to a driver. This can be done using signals developed by the control logic and the pull-down device MNDN.

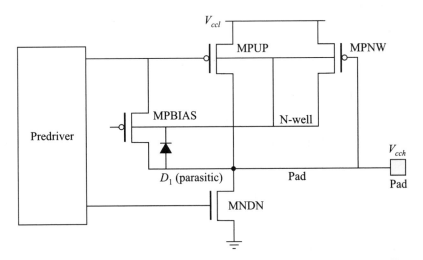

Figure 7-16. Self-biased N-well implementation of mixed-voltage output buffer.

The N-well switching action can prevent oxide damage and can dramatically reduce the leakage current. However, ESD protection is still not implemented and has to be added. For ESD protection of the buffer at least two solutions exist. One solution is to implement a diode chain, as shown earlier in the open drain case. The other solution involves the creation of a special ESD bus, which will be discussed in Section 7.4.

The high I/O and low-voltage core scenario was discussed, but now consider a high-voltage core and a low-voltage I/O. When the core is at a higher voltage than the I/O, then an NMOS pull-up device can be used in a parallel with a PMOS, as shown in Figure 7-17. This scheme can save area in some situations and reduce the pad capacitance, enabling faster I/O. Consider the case where an I/O is 2.5 V ($\pm$10%) and the core is 5 V ($\pm$10%) and a $|V_{tn}| = |V_{tp}| = 1$ V (20% of V_{cc}). Also assume that the driver will launch a wave into the transmission line equal to half the I/O voltage. If an NMOS pull-up is employed, it will have a V_{gs} drive of 4.75 V $-$ 1.375 V $=$ 3.375 V, which compared to the 2.25-V drive available to a PMOS device. In such a case, comparing the worst-case driving capability for an NMOS (low core voltage, high I/O voltage) to the PMOS, (low I/O voltage) the area saving is (assuming linear currents), the worst-case voltage and the electron mobility is twice the hole mobility: $(((3.375 - 1)/(2.25 - 1.0)) \times 2 = 3.8\times)$, which is a large savings. In addition, when the reflected wave arrives, it will create a V_{ds} of 0 V for the NMOS, and the NMOS will still be on (V_{gs} of 4.75 $-$ 2.75 $=$ 2 V, which is greater than a V_{tn} of 1 V). In reality the pull-up NMOS threshold voltage increases due to the body effect. Therefore, the gains will be reduced by some degree. However, consider a 5-V core, and a 3.75-V I/O. Here the savings are $(2.78 - 1)/(3.56 - 1) \times 2 = 1.4\times$. In addition, when the reflected wave arrives at the driver, the NMOS will turn off since V_{gs} (4.75 $-$ 3.93 $=$ 0.82 V) is smaller than the threshold voltage of 1 V. Therefore the trade-off of the NMOS pull-up

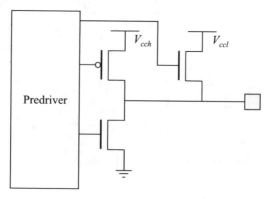

Figure 7-17. An NMOS pull-up can be considered when the core is at a higher voltage than the I/O voltage. The area advantage must be traded off with layout complexity.

device should weigh the area savings, the termination it provides, and the layout complexity.

7.2.3. Output Predrivers for High-Voltage I/O

The mixed-voltages I/Os have created special scenarios that impact the I/O predriver stage. An interesting problem is the design of a low-voltage predrive circuitry to control a high-voltage CMOS I/O buffer. The problem is due to the low voltage (core V_{cc}) at the PMOS driver gate when the predriver is driving "high"; the PMOS driver may not completely turn off and there is some crowbar current through the PMOS and NMOS transistors. Thus special circuit techniques are required. One such solution is shown in Figure 7-18.

The circuit is a simple voltage translator allowing the low-voltage core to control the high-voltage I/O PMOS [Declercq, 1993]. The idea is to implement the control using NMOS pull-down transistors and then have a cross-coupled structure and pull the I/O PMOS gate to a high. To drive the MPIO, the V_{in} goes high, which turns on MN_1 and pulls the PMOS gate low (V_{pbias} level). The low voltage should be kept at a level other than zero to prevent stressing the predriver PMOS devices. This can be supplied externally or generated on-chip. For turning off the PMOS MPIO the V_{in} is set low. This turns off MN_1 and

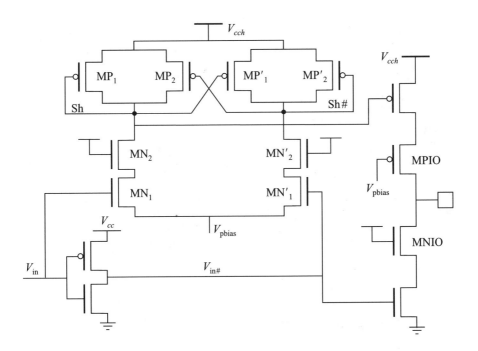

Figure 7-18. Simple voltage shifter and predriver for CMOS [Declercq, 1993]

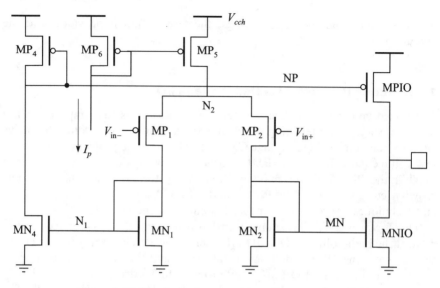

Figure 7-19. Analog voltage translator [Declercq, 1993].

turns on MN_1'. This in turn pulls the gate of MP_2 low, which charges the gate of MPIO to a V_{cch} level, thus turning it off.

A similar translator that has a differential input can be constructed, as shown in Figure 7-19. Here, a current mirror MP_6 is used to reflect into MP_5. Depending on whether V_{in+} or V_{in-} is higher, more current is steered into MP_2 or MP_1, respectively. If all the current is steered into MN_2, the node voltage MN rises, which turns on MNIO. To have substantially large voltage at MN, the current through MP_5 and the size of MN_2 should be chosen such that MN is near V_{cc} when all the current is steered into MN_2. This will allow MNIO to be smaller.

Similarly, when V_{in-} is lower, MP_1 steers all the current, which in turn biases N_1 and turns MN_4 on. As earlier, the sizes of MN_4 and MP_4 have to be chosen such that NP is set to near the PMOS oxide breakdown, which in turn reduces the size of MPIO. In this scheme, the current at the buffer is also linked to I_p in a ratio of device widths (MNIO : MN_2 and MPIO : MP_4).

In the previous sections the effect of mixed voltage on the input and output buffer designs have been discussed. The mixed voltage also has an impact on the ESD power supply coupling choices. These will be examined in subsequent sections.

7.3. EFFECT ON POWER SUPPLY COUPLING DIODES

One method to design a low-voltage system is to retain the core at the nominal technology voltage but operate the periphery at the high voltage. Thus the external system sees a consistent high-voltage I/O. This effectively moves the problem from say input leakage to that of leaking supply coupling diodes. This is illustrated in Figure 7-20. In this example a 5-V I/O and a 3.3-V core is being used in the chip. A simple diode chain may not be able to sustain the extreme voltages and temperatures without considerable leakage. Here again the leakage can be reduced using the techniques developed in Chapter 2 using biasing resistors. The PMOS transistors (acting as resistors) are used to supply the leakage current to the discrete taps in the diode chain. Since these resistors have to be in the megaohm range, they are therefore long. Now instead of tying the PMOS (resistor) gates to a low voltage (V_{ss}), an intermediate voltage (e.g., V_{tm}) will lower the drive strength of the PMOS bias transistors considerably, leading to smaller length resistors. This is shown in Figure 7-20 where the resistor PMOS gates are referenced to V_{cc} and not to V_{ss}. The reference to V_{cc} also avoids oxide reliability issues as the gate–bulk stress is reduced. This is termed a cladded diode.

Figure 7-21 shows that in the chip operating temperatures of 80–100°C the leakage can be reduced from a factor of 2–5× and greater when further optimized. At the higher temperatures (> 110°C) no difference is noted in what is measured and predicted (no biasing). This means that the PMOS resistor current is not sufficient to meet the leakage current requirement of the diodes; therefore, the current starts trickling in through all the diode stages, entailing with it the corresponding Darlington gain. Another factor at elevated temperatures is that the diode forward voltage decreases. Thus the 800 mV across each diode pair starts turning them on. The only solution to this issue then is to increase the number of diode stages.

An alternative method to design this cladded diode is to calculate the N-well leakage at a design temperature using either experimental or theoretical values. Then assume that the 2.5 V (5.5 V − 3.0 V) is dropped equally among the three segments of the diode chain. This fixes V_{ds} across each transistor. Knowing the current required by the N-well, V_{ds} and V_{gs} across each transistor, theoretical or simulated sizes can be calculated. These designs are fairly easy, and a simple experimental spread can lead to the right leakage and transistor size trade off.

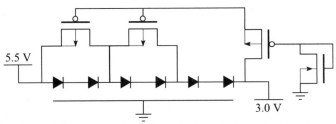

Figure 7-20. Cladded bias network implemented in P-channel FETs for a six-diode mixed power supply clamping string intended for process limited to 3.6 V across gate oxides.

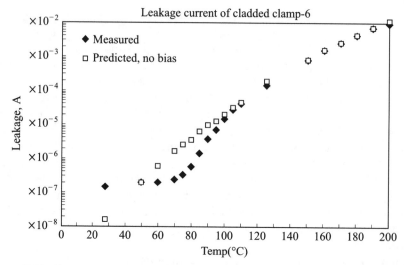

Figure 7-21. Six-stage P-channel cladded diode string current vs. temperature for Figure 7-20 compared with unbiased current as calculated from individual cell data.

7.4. SEPARATE ESD BUS

When large differences in voltages exist between a pad and V_{cc}, a large number of cascaded ESD diodes are required. The pad-to-V_{cc} diode requirement can be drastically reduced if an intermediate supply is introduced in place of V_{cc}. However, now the intermediate supply and V_{cc} need to be clamped and a large number of cascaded diodes may again be required. A solution is to provide an independent ESD supply, as shown in Figure 7-22. The ESD reference supply either can be tapped from outside the chip or can be floated and then referenced to a known voltage (in this case V_{cc}) [Worley, 1995]. The external ESD reference has itself to be clamped, and a convenient method to clamp this node is to use a cantilevered diode clamp or an NMOS clamp.

This partitioning leads to a lower capacitance at the input node, which contributes to faster transitions. The separate power bus results in a significantly lower area requirement for the diode. This can be seen in Figures 7-22c, d. The single diode reduces the buffer height significantly. The additional diodes required to clamp the ESD reference bus to V_{cc} are shared, and this can be conveniently placed in the relatively free power pad areas. If the V_{esd} bus is clamped to ground using MOS clamps or cantilevered clamps, these too can be placed easily in the power pad areas.

Sometimes the high voltage of the ESD reference bus may not be tolerable by the ordinary cantilevered clamp built in a low-voltage technology. In such a case the cascade PMOS clamp design techniques used in the output drivers can be used [Worley, 1995]. By placing the voltage clamping transistor, the gate-to-

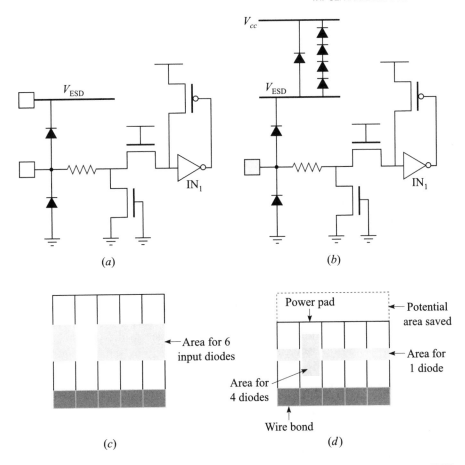

Figure 7-22. Two schemes to reduce diode stack-up by providing (*a*) a separate ESD rail connected externally [Voldman, 1995] and (*b*) an independent ESD rail that is self-biased [Voldman, 1994]. (*c*) Chip layout with six-diode chain per input buffer and (*d*) chip layout with separated ESD bus using only one diode between pad, ESD reference bus, and four diodes shared between ESD reference and V_{cc}. Additional clamp diodes between the ESD reference bus and V_{cc} are shown utilizing power pad area.

bulk-oxide fields are reduced to a range tolerable by the low-voltage technology. If a cantilevered diode is desired as a clamp, suitable biasing arrangements have to be made to the termination. An example is shown in Figure 7-23 [Maloney, 1997]. The diode biasing transistors MP_1, MP_2, and MP_3 are protected by the combination of transistor biasing transistors MP_5 and MP_6. The PMOSFETs MP_5 and MP_6 are long channel devices such that they provide enough leakage such that node N_{cap} is close to V_{cc} but long enough to reduce the leakage from V_{cc} to V_{ss} and to the low-microampere range. This prevents the transistor MP_4 from being overstressed. If the diode chain cut-in voltage is

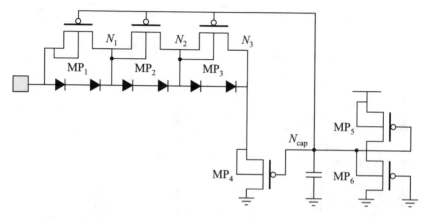

Figure 7-23. Cantilevered diode adapted for high-voltage operation.

assumed to be 3 V at high temperatures, then during circuit operation the cantilever diode will not turn on for $3\,V + V_{cc} + V_{tp}$ volts.

The discussion on the high-voltage drive or the tolerant drive was achieved by oxide stress and hot-electron management primarily by self-biasing the N-well and cascading the drivers. For the ESD protection of such drivers an independent ESD reference power can be utilized. This avoids long diode chains in the self-biased N-well mixed-voltage output buffer when an independent core clamp (i.e., cantilevered diode or NMOS clamp) is used. This is shown in Figure 7-24. The ESD reference voltage brought in externally is directly clamped to V_{ss} using a cantilevered diode. This concept can be extended to a completely independent ESD reference supply that is self-generated on-chip using a charge pump. Usually the current requirement is small in such reference buses with small leakage currents; thus only a small charge pump is required (see Figure 7-25).

Since these supplies have very limited current-handling capabilities when the chip is on, they will provide no clamping for large I/O overshoots. If the diode is to be used as a signal clamp also, then the ESD reference bus and the V_{esdref} must be pinned externally multiply to lower the inductance compatible with the di/dt injected noise. If the ESD reference bus is not suitable powered externally then it will act as a peak detection circuit.

A simple charge pump construction is shown in Figure 7-26. The action of the pump is as follows. When the clock is high, the NMOS MN_1 grounds the C_1 terminal, effectively charging the node to V_{cc} less a diode drop. When the clock goes low, the PMOS P_1 turns on, raising the potential of the n_1 to $V_{cc} + (V_{cc} - V_{diode})$. A similar action is then repeated at the clock number signal using transistors MN_2, MP_2, and Q_2. Thus node V_{cc2} charges to a voltage higher than the V_{cc} level.

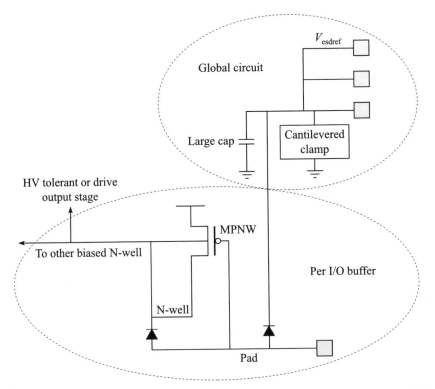

Figure 7-24. Simplified ESD protection scheme using Voldman's basic self-biased N-well structure. The V_{esdref} can be multiply bonded out so it is compatible with the I/O signal clamping requirements.

7.5. BACK BIASING

A crucial issue in ESD design is the biasing of unpowered chips connected to a live system [Voldman, 1994; Tong, 1996] through ESD circuits. A scenario is shown in Figure 7-27. A driver drives a line connected to two components of a system, one powered on and the other off. The signal line forward biases the input protection diodes of the powered-off system. The off chip may cause sufficient leakage that the proper signal voltages are not reached by the signal line. Consequently an I/O failure results.

A simple placement of a cantilevered diode or MOS clamp in the V_{cc} power supply may not solve the problem. The V_{cc} core voltage may rise just sufficiently to cause leakage in the core circuits. The core voltage may not rise sufficiently to turn on any core clamps, but these usually are not the issue. Therefore any direct pad-to-V_{cc} coupling through a diode will enable leakage into the powered-off system. One solution therefore is to remove the pad-to-V_{cc} coupling diode and replace the ESD protection with some other scheme

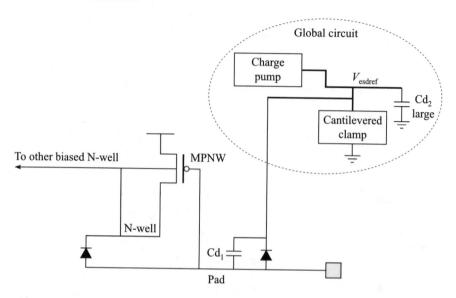

Figure 7-25. Completely isolated ESD reference bus. For this case V_{esdref} acts as an isolated supply.

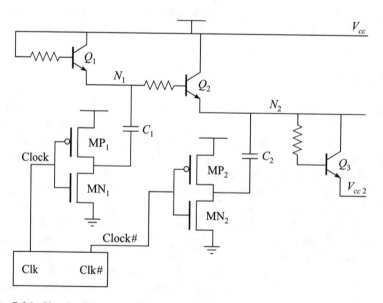

Figure 7-26. Simple charge pump that can be used to generate a separate ESD reference supply [Okamura, 1996].

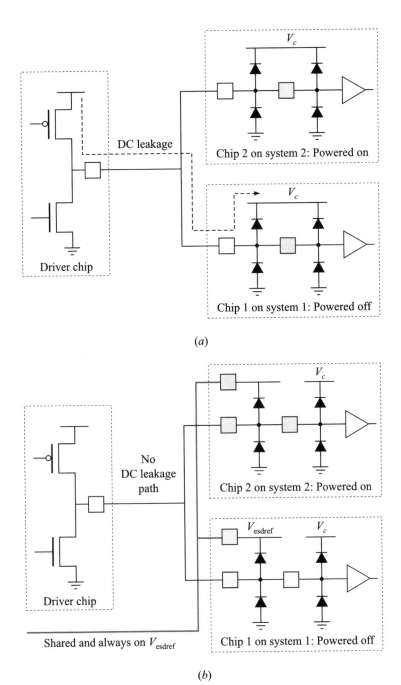

Figure 7-27. System issue due to ESD diodes: (*a*) powered off-chip can sink current and may not allow correct signal levels (*b*) by breaking the V_c into V_c for the core and V_{esdref} reference for the I/O; thus isolation can be obtained.

(self-protecting pad using NMOS, SCR, protection, etc.). These are usually difficult to design.

In this case, a solution is to have an independent ESD reference supply not coupled to V_{cc} and ESD protected by a cantilevered diode or an MOS clamp. Since the leakage path is only the clamp that will turn off as the capacitance of the RC sense element charges up, a self-isolating supply is constructed. This enables no "backbiasing" and easier system design. This is shown in option (b) of Figure 7-27. If a shared and always powered-on V_{esdref} is used, then this backbiasing can effectively be controlled.

7.6. PROCESS MODIFICATIONS TO SUPPORT HIGH VOLTAGES

Sections 7.1–7.5 have examined circuit methods used to design mixed-voltage interfaces. When the process voltage difference is within one generation (e.g., 5 V and 3.3 V) usually circuit design can be effectively implemented. However, if the process voltage difference is higher than one generation, then it is extremely complicated and expensive to have a mixed-voltage interface. In such cases technology modification may be required. Two methods are to increase the gate oxide breakdown voltage (thicker oxide) and to increase the drain-to-source-voltage capability by drain engineering (double-diffused MOS).

For high-frequency and high-voltage applications double-diffused MOS (DMOS) has been utilized [Poocha, 1974; Sanchez, 1989; Pierret, 1990]. The DMOS structure is shown in Figure 7-28. These transistors can also serve well as output buffers. The channel length can be fairly small and depends only on the diffusion doping profiles rather than solely on the lithography capability. However, lithography will determine the registration to the channel, thus the gate-to-source and drain-to-gate parasitic capacitance.

A way to provide voltage-tolerant I/O is to increase the oxide thickness of the I/O, as depicted in Figure 7-29. This is an added process step as the core circuits still work on the regular process oxide and only the I/O is touched. In

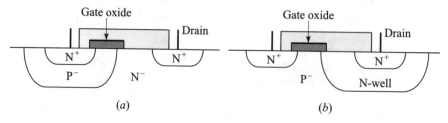

(a)   (b)

Figure 7-28. A DMOS structure in a (a) N-epi CMOS technology. Since the drain is in an N-well, the drain junction capacitance is small and the drain is high voltage tolerant (b) in a P-epi technology.

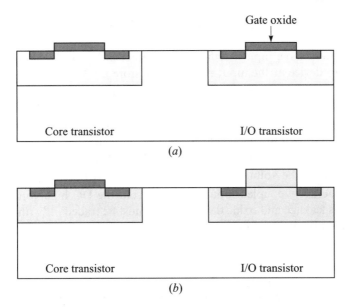

Figure 7-29. (*a*) Normal process utilizes the same gate oxide for the core and I/O circuits. (*b*) For very high voltage applications a thick oxide on the I/O may be used.

some processes such as flash memory, dual oxides are regularly supported [Otsuka, 1997]. The impacts of the increased oxide are higher breakdown voltage and more hot-electron-resistant devices. Simultaneously, the threshold voltage increases and G_m decreases, thus making the devices large [Declercq, 1993]. However, they may still be smaller than the cascading options.

Since the number of process steps increases, manufacturing losses will increase. In addition, the wafer throughput will be slower through the fabrication line, thereby increasing the cost of each chip. However, in some cases this may be the only viable option (i.e., I/O voltages being much higher than those supported by the core technology, e.g. 3.3 V I/O on a 1-V technology).

The thick-oxide transistors can also be used directly in input buffers. The need for a pass gate to limit the high-voltage swing is removed, and this results in cleaner input signals.

7.7. SUMMARY

Ever since the first DRAM lowered the voltage from a standard 5-V power supply to 3.3 V there has been a continuing lowering of the power supply voltages. This trend will certainly continue (at least to the 1-V range). However, this has led to several issues that need resolution.

1. Input buffer design has been affected:

   - The low-voltage input circuitry needs either a differential amplifier or a pull-up device at the input receiver to restore a CMOS level.
   - The high-voltage input usually requires additional diodes to prevent a leakage path between the pad and V_{cc} and also a pass transistor to clamp the high-input level seen by the receiver. This pass transistor adds delays, skews, and asymmetry to the received waveform.
   - To attenuate a high-voltage input, a resistor divider can be employed to bring down the voltage to the level supported by the process level.
   - A capacitively coupled input can also avoid the DC high-voltage issues such as leakage and oxide damage. This must be traded off with increased complexity in circuit design.

2. The output driver becomes involved:

   - For high-voltage-tolerant open-drain drivers the NMOS device must be cascaded. The top device must be biased (usually at V_{cc}) to increase the size and delay.
   - For a CMOS high-voltage-tolerant driver, the PMOS must also be cascaded. To prevent a leakage path from the pad to the V_{cc}, the N-well must be floated when the driver is inactive.
   - For a CMOS high-voltage driver, the PMOS must also be cascaded but the N-well does not need to be floated.
   - The output predriver for high-voltage drivers needs to be modified so that it can turn off the PMOS device.

3. The power supply issues increase:

   - The mixed-voltage rails in a chip need protection. If they are coupled to each other using a diode chain, then a longer chain is required to allow for the worst-case voltage differential and sufficient noise isolation.
   - To reduce the number of diodes in a high-voltage input scenario, an independent high-voltage ESD bus can be utilized. This allows only one diode between the pad and the power supply, resulting in area and capacitive loading savings. The independent supply can be either supplied externally or generated internally using charge pumps.
   - Process modifications in terms of thicker oxides and DMOS output transistors may be required if the core and I/O voltages differ by more than one process generation.

REFERENCES

[Chen, 1996] M. J. Chen, J. S. Ho, and T. H. Huang, "Dependence of Current Match on Back Gate Bias in Weakly Inverted MOS Transistors and Its Modeling," *IEEE JSSC*, **31** (2), 1996, p. 259.

[DeClercq, 1993] M. J. Declercq, M. Schubert, and F. Clement, "5V to 75V CMOS Output Interface Circuits," Digest of Technical Papers, *ISSCC*, 1993, p. 162.

[Gabara, 1997] T. Gabara and W. C. Fischer, "Capacitive Coupling and Quantized Feedback Applied to Conventional CMOS Technology," *IEEE JSSC*, **32** (3), 1997, 419.

[Gardner, 1980] F. M. Gardner, "Charge-Pump Phase Lock Loops," *IEEE Trans. Commun.*, **COM-28** (11), 1980, 32.

[Greenhill, 1997] D. Greenhill et al., "A 330 MHz 4-way Superscalar Microprocessor," *ISSCC*, 1997, p. 166.

[Hu, 1994] C. Hu, "Low-Voltage CMOS Device Scaling," *Proc. ISSCC*, 1994, p. 87.

[Hulett, 1990] T. Hulett, "On Chip Protection of High Density NMOS Devices," *Proc. EOS/ESD Symp*, 1981, p. 90.

[Krakauer, 1992] D. Krakauer and K. Mistry, "ESD Protection in a 3.3V Sub-Micron Silicided CMOS Technology," *Proc. EOS/ESD Symp*, 1992, p. 250.

[Maloney, 1997] T. J. Maloney, K. Parat, N. K. Clark, and A. Darwish, "Protection of High Voltage Power and Programming Pins," *Proc. EOS/ESD Symp*, 1997, p. 246.

[Okamura, 1996] Okamura, *JSSC*, 1996, p. 84.

[Otsuka, 1997] N. Otsuka and M. A. Horowitz, "Circuit Techniques for 1.5V Power Supply Flash Memory," *IEEE JSSC*, **32** (8), 1997, p. 1217.

[Pierret, 1990] R. F. Pierret, *Field Effect Devices,* Modular Series on Solid State Devices, 2nd ed., Addison-Wesley, Reading, MA, 1990, p. 136.

[Pocha, 1974] M. D. Pocha, A. G. Gonzalez, and R. W. Dutton, "Threshold Voltage Controllability in Double-Diffused-MOS Transistors," *IEEE Trans. Electron. Dev.*, **Ed-21** (12), 1974, p. 778.

[Prince, 1992] B. Prince and R. H. W. Salters, "ICs Going on a 3-V Diet," *IEEE Spect.*, 1992, May, p. 22.

[Sanchez, 1989] J. J. Sanchez, K. K. Hsueh, and T. A. Demassa, "Drain Engineered Hot-Electron Resistant Device Structures: A Review," *IEEE Trans. Electron Devices*, **36**, 1989, p. 1125.

[Tong, 1996] M. Tong, R. Gauthier, and V. Gross, "Study of Gated PNP as an ESD Protection Device for Mixed-Voltage and Hot-Pluggable Circuit Applications," *Proc. EOS/ESD Symp.*, 1996, p. 280.

[Voldman, 1994] S. H. Voldman, "ESD Protection in a Mixed Voltage and Multi-Rail Disconnected Power Grid Environment in 0.50 and 0.25 μm Channel Length CMOS Technologies," *EOS/ESD Symp.*, 1994, p. 125.

[Voldman, 1995] S. H. Voldman, G. Gerosa, V. P. Gross, N. Dickson, S. Furkay, and J. Slinkman, "Analysis of Snubber-Clamped Diode-String Mixed Voltage Interface ESD Protection Network for Advanced Microprocessors," *Proc. EOS/ESD Symp*, 1995, p. 43.

[Waizman, 1995] A. Waizman and N. Shanan, Delay Line Loop for On-Chip Clock Synthesis with Zero Skew and 50% Duty Cycle, U.S. Pat. 5,410,263, April 25, 1995.

[Worley, 1995] E. R. Worley, R. Gupta, B. Jones, R. Kjar, C. Nguyen, and M. Tennyson, "Sub-Micron Chip ESD Protection Schemes Which Avoid Avalanching Junctions," *Proc. EOS/ESD*, 1995, p. 13.

[Yoo, 1995] C. Yoo, M.-K. Kim, and W. Kim, "A Static Power Saving TTL-to-CMOS Input Buffer," *IEEE JSC*, **30**(5), 1995, p. 616.

[Young, 1992] I. Young, J. K. Greason, J. E. Smith, and K. L. Wong, "A PLL Clock Generator with 5 to 100 MHz Lock Range for Microprocessors," *IEEE ISSCC*, 1992, p. 50.

[Zanders, 1991] G. V. Zanders, ESD Circuit Which Exceeds Power Supplies in Normal Operations, U.S. Pat. 5,032,742, July 16, 1991.

[Zupac, 1992] D. Zupac, D. Pote, R. D. Schrimpf, and K. F. Galloway, "Annealing of ESD-Induced Damage in Power MOSFET's," *Proc. EOS/ESD Symp.*, 1992, p. 121.

CHAPTER 8

ESD RELIABILITY MEASUREMENT AND FAILURE ANALYSIS BASICS

The ESD and I/O design aspects and how they interrelate have been discussed in previous chapters. In this chapter the discussion will focus on

- ESD and related stress and stressing instrumentation and
- failure analysis of EOS/ESD-related defects.

First simple ESD instrumentation will be discussed followed by a brief section on failure analysis techniques.

8.1. SIMPLE ESD INSTRUMENTATION

Sophisticated ESD stressing tools are available from a number of equipment suppliers as Oryx and Keytek [1990]. These tools are fairly sophisticated and automated and aimed primarily at a manufacturing environment. These tools attempt to replicate the ESD models discussed in Chapter 1. A great deal of attention is paid with respect to minimizing the circuit parasitics that can invalidate the stressing waveform. For the laboratory these tools may be too elaborate, too expensive, too large, or not available at all. In such a case, simple stressing tools can be constructed. In this section three elementary but extremely useful instruments will be described. The stress generation apparatuses discussed here are the

- human body model,
- charged device model, and
- transmission line.

In addition to the above instruments, there are a host of tools used for ESD characterization, such as a charge plate monitor, an electrostatic field meter, a residual voltmeter, and a surface resistivity meter, and a good starting reference is Dangelmayer [1990, p. 101].

8.1.1. HBM Instrumentation

The HBM model is fairly simple to implement as it is not very sensitive to electrical parasitics such as stray capacitance and excess lead inductance. A schematic showing a simple construction is shown in Figure 8-1. The design can be broken down into a charging circuit and a discharging circuit. A high-voltage source can trickle charge the capacitor C_2 (100 pF) through the relay switch S_1. Once the charging is complete, the chip is zapped using S_2, which is mutually exclusive to the S_1 switch. The 100-pF capacitor then discharges through the 1500-Ω resistor into the chip [Antinone, 1986].

The grounded pin can be shorted to the ground plane with a small wire to reduce inductance. In the HBM test, the inductance is not a crucial factor because of large rise times and the 1500-Ω resistor in series. Consider the HBM model with a frequency component of 2.1 MHz; the inductance impedance that will equal the resistance of the HBM model (1500 Ω) is 10 μH. This shows that tens of nanohenrys of parasitic inductance will not significantly perturb the HBM test. Inductance of $\sim$10 nH can therefore be easily tolerated, which means the inductive loop between the pin and the V_{ss} pin can be on the order of a few inches. This is handy in making the wires flexible with pogo pin ends (usually) to attach to pins of the chip. The loose wires allow easier testing and maneuverability over the chip.

Human body model testing is conducted in accordance with the ESD association specification [HBM, 1993]:

- Each I/O pin should be stressed against each power supply pin. It is permissible to short all the power supply pins together. In automated testers this is easily done, but in manual testers, it is not convenient to

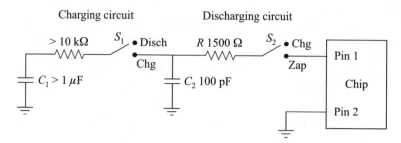

Figure 8-1. Simple HBM box constructed using relays, resistors, and capacitors.

short all the tester pins. In this case, only one power pin can be connected.

- Each power supply can be stressed with respect to other power supplies.
- Three repeated zaps in sequence are required, and there should be at least a 1-s interval between consecutive zaps. However, this time period is being reduced to 300 ms.

Today HBM is not considered as threatening as, say, in the 1980s, primarily because of decreased manual handling of parts, enhanced understanding of protection circuits, better ESD control, implementation, and management [Danglemayer, 1990, p. 64]. However, another threat has come up due to increased automation, and this is the machine charging the parts and then discharging them, as reflected in the charged-device model.

8.1.2. CDM Instrumentation

The CDM event is considerably faster than the HBM event, making it harder to model and sensitive to parasitics. Currently this has resulted in two implementations of the test: the field-induced tester [Renninger, 1991; Oryx] and the socketed test [Keytek, 1990]. The field-induced test works on the principle of charging the chip and then moving a grounded pin near the chip's pin. This results in an air discharge (noncontact) depending on the voltage and the pin distance. This test closely resembles the actual ESD zap experienced by the chip. However, the spread of current–time distribution in this test is not well controlled and depends significantly on the rate of approach and on the environment. The socketed version is similar to the HBM construction. Here the component is placed in a socket, being charged by a field plate (similar to the nonsocketed version), but the discharge is again controlled by a relay. The relay environment is constant but different from reality.

The socketed version does not suffer from a rate of approach issues and has little environmental effect. However, it does not truly replicate a CDM zap and only provides a close equivalence. Further, the field-induced tester can only be used in CDM zapping. The socketed versions can easily swap between HBM and CDM testing.

In the laboratory, a simple CDM tester can be constructed using commonly found materials; the physical construction is shown in Figure 8-2 and the electrical schematic in Figure 8-3. It consists of a box, two small wires (low inductance) a clamp to physically hold the device, and an electrical apparatus such as a relay and a resistor. In this box, the chip is placed in a dead bug fashion (pins pointing up) over the ground plane. The V_{ss} pin is shorted to the charging voltage. Then the charging plate is slowly raised (i.e., trickle charge due to the high charging resistance) to a test potential. Then the charged capacitor (between the chip and the now grounded charging sheet) is discharged through a pin. The discharge pin inductive impedance should be

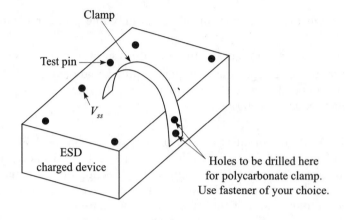

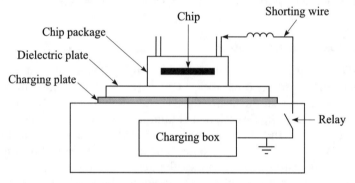

Figure 8-2. Laboratory CDM box: (*a*) physical construction; (*b*) cross section. One key parameter is to watch for parasitics. The relay should have the least capacitance and the wires the least inductance.

much smaller than the arc resistance due to a CDM zap. If the arc is assumed to have an average resistance of 50 Ω [Maloney, 1992, 1993] and a bandwidth of 1100 MHz [Amerasekera, 1995], then the inductance has to be much less than 8 nH.

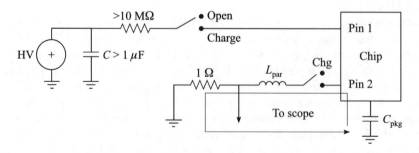

Figure 8-3. Simplified electrical model of CDM box.

The simplified box is shown in more detail in Figure 8-4. This shows the exact hook-up of the relays, the switches for the curve tracer (analysis), and the resistor to enable scope measurements.

The previous sections have examined equipment that tries to simulate an ESD zap model directly using electronic components. Another method to do ESD stressing is to use a simpler stress tool and then extrapolate that data to an ESD regime. This approach is discussed next.

8.1.3. Transmission Line Stress Box

As mentioned earlier, the CDM box is fairly parasitic sensitive. The repeatability of these measurements depends on the exact fixture and placement of the wires. A much easier method to characterize, model, and stress devices is to use a transmission line as a current source, as shown in Figures 8-5 and 8-6 [Maloney, 1985, 1990]. The ease of this method has allowed it to be widely

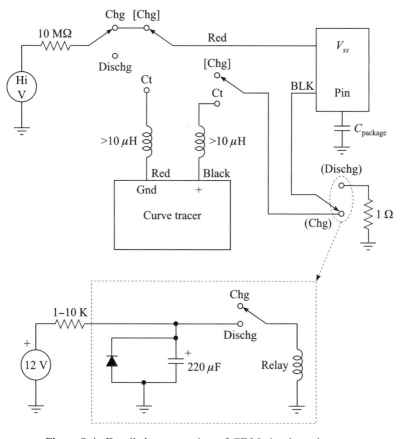

Figure 8-4. Detailed construction of CDM simulator box.

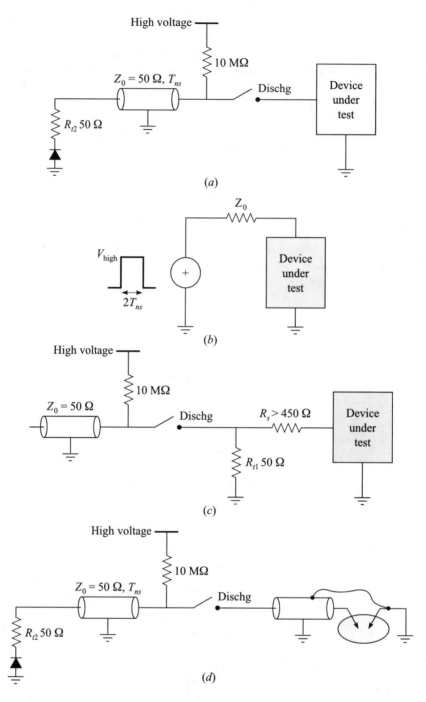

Figure 8-5. (*a*) Simple transmission line apparatus and (*b*) its equivalent circuit. (*c*) High-impedance transmission line pulser can be constructed by placing a resistance in series. (*d*) Setup can also be modified to suit a wafer level characterization of devices [Maloney, 1985, 1990].

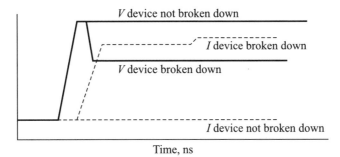

Figure 8-6. Current and Voltage waveforms experienced by device under test.

accepted [Amerasekera, 1992; Diaz, 1992, 1994b; Polgreen, 1992]. In this method, a transmission line is first charged up to a known voltage. This voltage corresponds to the desired current in the line. For example, a transmission line of 50 Ω impedance with a series resistor of 475 Ω can deliver current of 2 mA if it is charged to 1000 V and then the end is shorted to ground. The shorting (or low impedance) occurs when the device under test breaks down. Otherwise, the line is terminated with a matching resistor to minimize reflections. The total amount of charge depends on the length of the transmission line. The transmission line has increasingly become popular as it is well calibrated and easy to use. The parasitics can be kept small and repeatability is enhanced. In some cases deliberate parasitics have been placed to degrade the edge. Also increasingly models correlating the transmission line with CDM and HBM are being developed [Pierce, 1988; Beebe, 1996; Gieser, 1996].

The voltage can be monitored on the scope across the termination resistor. Using this, the current and the voltage can be calculated in the device, enabling accurate estimation of the power delivered in the device.

Sections 8.1.2 and 8.1.3 dealt with the construction of the EOS or ESD stress equipment. There are other tests for reliability that, although not directly related to ESD, do significantly impact ESD protection. One such test is the burn-in electrical and temperature overstress aimed at screening out weak parts. Although this is not a direct ESD stress, it can cause EOS damage to poorly designed circuits. In addition, the designs of diodes and clamps are significantly affected due to higher temperatures and voltages. These effects are considered next.

8.1.4. Burn-in Testing

The burn-in screening is an important method to screen out weak/defective parts quickly before they are used in systems [Intel, 1991]. The basic assumption of burn-in screening is that the parts will show a bathtub reliability curve. This tub, shown in Figure 8-7, plots the failures as a function of time. The early failures are due to manufacturing defects and weakness in the circuits. These

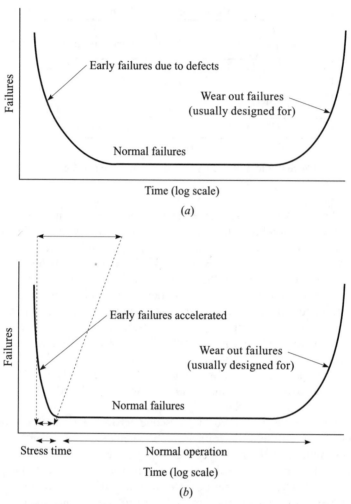

Figure 8-7. Bathtub curve showing failures as (*a*) a function of time with normal operation and (*v*) with burn-in acceleration time followed by normal operation.

may be due to poor metallizations, vias, oxides, contamination, assembly, and other reasons. Once these early defects are screened out, the rest of the parts show a steady but significantly lower failure rate. This is the part's normal operating life. After the part ages, the failure rate again rises sharply. This is due to the wear-out mechanisms (i.e. hot-electron, electromigration junction defects, oxide wear-out, etc.).

Early failures can take weeks to fail. Thus under normal circumstances weeding out the early defects would tie up equipment for extended periods. This is clearly unacceptable. Thus some method to accelerate the screening is needed. Here another assumption relating the wear-out rate to various stress parameters is required and is usually well justified. For example, if the voltage and temperature are increased, it is found that parts wear out quicker. The key is to relate the acceleration factors to the wear-out rate. The acceleration factors are usually exponentially related to the increased voltages and temperatures. Thus, 2 weeks of stress screening at normal voltages can be achieved in sufficient hours.

The acceleration factors are achieved using higher voltages than in a normal operating circuit. The voltage acceleration factor (VAF) is given as

$$\text{VAF} = e^{[C(V_s - V_0)]} \tag{8-1}$$

where V_s and V_0 are the stress and normal operating voltages and C is a constant determined by the technology.

Similarly, the increased temperature enhances failures of all Arrhenius-related defects. The acceleration can be given by

$$\frac{t_1}{t_2} = \exp\left[\frac{E_a}{k}\left(\frac{1}{T_1} - \frac{1}{T_2}\right)\right] \tag{8-2}$$

where t_1 and t_2 are the mean times to failure (MTTF) and T_1 and T_2 are the temperatures (in Kelvin); E_a is the activation energy of the defect and k is the Boltzmann constant, 8.62×10^{-5} eV/K.

For burn-in testing components are placed in a burn-in oven. Typical burn-in temperatures can be as high as 160°C with voltages up to 20% of nominal. This has a direct impact in some ESD protection schemes. For example, these imply high leakage for diode chains unless proper precautions (i.e., leakage suppressors using biasing techniques) have been taken. Under such circumstances it is possible to have the diode chain experience thermal runaway, which will result in permanent damage to the chip. These voltage and temperature considerations increase the number of diodes in a string, which in turn increase their area and their forward voltage. For MOS-based clamps the burn-in is not such a severe concern because these only experience higher leakage. To summarize, the burn-in may impose stricter design constraints for EOS/ESD circuits than normally encountered during normal circuit operation.

Once the devices have been stressed, they are screened for functionality. This can be done using either electrical or optical means. Currently smarter burn-in boards can do simple tests while in the burn-in chamber itself, and separate testing may not be required.

8.2. FAILURE ANALYSIS

Failure analysis as applied to ESD is that phase of chip product development in which failed parts are diagnosed with the intention to correct the problem or improve their robustness. As can be seen in Figure 8-8, it is one of the last

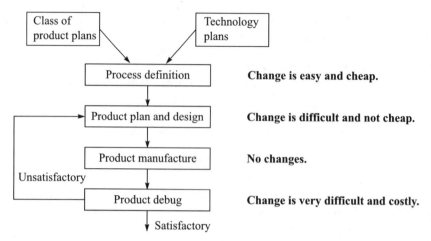

Figure 8-8. Product development cycle showing flexibility and relative cost of implementing changes.

phases in product development. Due to its late nature, it is very inflexible and expensive. A major problem at this stage results in an expensive and time-consuming cycle. The diagnosis can be difficult and sometimes very involved. When rare and hard-to-duplicate defects are encountered, a correct diagnosis is not even a certainty.

Inevitably most complex products need a debug phase. A significant portion of failure-analyzed parts are due to EOS and ESD damages [Diaz, 1992; Euzent, 1991]. A failure analysis study by Diaz [1992] indicated that as a percentage of total failures the EOS/ESD failures ranged 3–26% and failure analyses requested by customers (returned from the field) accounted for 5–70% of total requests. This indicates that there are significantly higher field failures attributed to ESD than other sources. Due to their late nature of detection, as can be seen in Figure 8-8, these failures are the most expensive to rectify. Keeping in view the expense and the few options available resulting in significant delays, it is imperative that ESD design be done right the first time and to effectively screen out any defects that may have gone unnoticed. The screening can be done either electrically or microscopically.

Failure analysis techniques favor three features: nondestructiveness, ease, and resolution of the fault. The most favored ones are attempted first, followed by other techniques. Euzent [1991] suggests a failure analysis flow that meets this general concept. In Euzent's proposal, first a visual inspection can be done using a simple microscope and scanning electron microscopy (SEM), which results in 10% resolution of the problems. This can be followed up with an electrical characterization using data sheets and parametric testers, which results in an additional 55% failure resolution. For the still unresolved problems, other nondestructive analyse need to be done, such as optical microscopy, emission microscopy, and hot-spot detection using liquid crystal or infrared cameras. These result in an additional 25% failure resolution. The total problem resolution is about 90%. For the 10% unresolved problems, the fault can be isolated to an electrical node with the aid of ion milling and testing using electron beam testers and microprobes. This in turn can be followed up by subsurface analysis. The subsurface analysis may require etching off the interlayer dielectric and other oxides to enable optical or SEM inspection on a layer-by-layer basis. This brings another 7% failure resolution. Finally, complex subsurface analyses such as precision cross sections, high-resolution SEM, and other microscopic techniques can be performed.

As suggested earlier, after a quick optical examination, usually electrical methods of diagnosis are easiest and quickest testing techniques. An example is presented here that demonstrates the efficiency of electrical testing and failure detection.

8.2.1. Electrical Evaluation

The standard method to detect ESD damage, especially at the manufacturing level, is electrical testing. The standards are specified in terms of electrical measurements as allowable leakage and leakage measurement conditions. Electrical methods are usually the quickest, cheapest, and to a degree most noninvasive, which explains their popularity [Euzent, 1991]. Parts that exceed the leakage specification are sent for further analysis to the laboratory.

Usual ESD damages are dead shorts or opens with extensive damage. These are fairly easy to detect. However, a number of issues arise in testing that can cause complications in interpretation of the electrical test data:

- cold annealing when a failed device recovers after some time,
- protection window gaps when ESD protection is effective for lower and higher stress voltages but fails at some intermediary stress level [Avery, 1991], and
- ESD hardening when one ESD zap causes the worst-case degradation as opposed to the conventional belief that larger number of zaps cause greater device degradation.

In addition, there may be some cryptic and hard-to-decipher damages. For example, how can we distinguish between a random and an ESD damaged oxide? Consider the typical gate oxide failure shown in Figure 8-9. When gate oxides are shorted due to the ESD energy, a polysilicon filament shorting the gate to the N-well or the substrate forms. The N^+ diffusion in a P substrate (NMOS) or P^+ diffusion in an N-well (PMOS) causes diode formation. This diode will rectify and have a temperature-dependent characteristic. If the N^+ is shorted to an N-substrate or N-well, a direct short is formed. No diode or temperature effects (except resistive) will be seen [Kim, 1992]. Thus, ESD-induced damage will create a PN junction or create a dead short. Both of

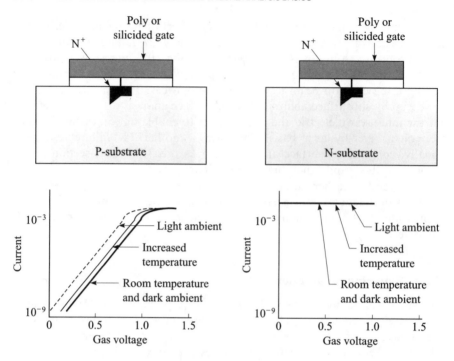

Figure 8-9. Model of observed leakage current due to ESD damage during water fabrication.

the above have significantly higher leakage compared to the leakage caused by random defects. The type of oxide damage mechanism that occurs due to EOS/ESD during manufacturing is becoming increasingly important as can be witnessed by research being conducted in the field [Park, 1996; Gabriel, 1996; McCarthy, 1990].

This electrical test can aid in identifying the problem, although the confirming test is still microscopy. If the leaking area shows surface pitting, then ESD damage can be confirmed. Random oxide defects usually will not show changes in leakage due to light and temperature effects. If ESD damage causes PN junction formation, the diode is greatly affected by the temperature and light. The high temperature and the presence of light both dramatically increase the leakage. The light and temperature can then be used to identify ESD damage in the gate oxide, and distinguish it from a random oxide defect.

There are various electrostatic exposures to a wafer under fabrication. Typical process steps such as ion implantation and plasma etching impinge ions onto the wafer. Therefore, the wafer is typically more exposed to positively charged than to negatively charged ions. When the device is overstressed during fabrication but not sufficiently to cause catastrophic damage, degradation in G_m and shifts in threshold voltage are also observed [Kim, 1992].

Consider the metal wires connected to the gate oxide that will charge in a plasma-rich environment and then in turn stress the gate. This metal acts as an antenna to attract the charges and conduct it to the gate. usually a maximum ratio of metal to gate area is established for a technology and this is strictly adhered to. If this ratio is exceeded, the charge picked up by the metal is sufficient to cause degradation of the oxide such that circuit operation or reliability is seriously affected [Intel, 1991]. If the metal area to the gate area is exceeded, diodes (N^+ in P substrate) or extra dummy gate capacitance have to be employed. The effects of the diode are shown in Figure 8-10. These help in charge sharing and provide a leakage path for the charge, thereby preventing oxide damage. The diode leakage is greatly increased when there are photons present in the process. This is generally true in plasma processes. Here, a small N^+ in a P-epi diode has sufficiently large photocurrent that it prevents damage even for very large antenna ratios [Park, 1996]. Protecting oxides with diodes was greatly aided by the fact that the oxide breakdown voltage was much larger than the junction breakdown voltage. However, the voltage difference is decreasing; therefore the effectiveness of the diodes will diminish.

The PMOS and NMOS show different levels of damage when the gate is positively biased. For the PMOS the surface potential at the silicon–oxide interface is determined by voltage sharing between the oxide and the depletion layer and is shown in Figure 8-11. This voltage sharing reduces the PMOS gate oxide stress; thus the PMOS shows minor threshold shifts due to processing. However, the NMOS voltage is dropped across the oxide due to the presence of an inversion layer directly under the gate at high positive gate voltages. Thus the NMOS shows larger threshold shifts (lowers) due to higher surface potential and fields at the surface. However, both the NMOS and PMOS experience G_m degradation.

When the charge on the gate is negative, both N and PMOS degrade equally. In the PMOS case, the P-epi/N-well diode is forward biased and the charges accumulate on the silicon-oxide interface, leading to a similar stress as

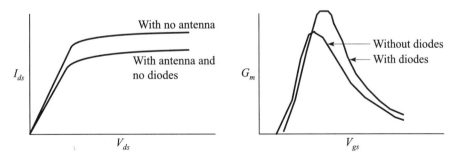

Figure 8-10. Effect of water fabrication stresses on unprotected MOS device suffering G_m and V_t degradation [Kim, 1992].

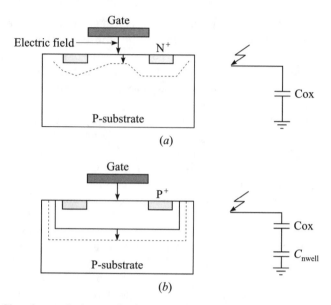

Figure 8-11. Charging and charge sharing at PMOS and NMOS gates when gate is positively charged. Location of electric field shown by arrows.

at the NMOS. In this case, the fields are about the same and both devices suffer similar damage.

It should also be noted that there are instances where electrical analysis is not sufficient to identify or detect a problem. Polyresistor damage is one such example [Amerasekera, 1995; Duvvury, 1988]. Here Duvvury shows physical damage to a polysilicon resistor, but it was electrically indistinguishable even after severe degradation. Only visually was this defect detected. Thus, electrical analysis can aid in quickly identifying problems, but it should be used as one of the tests in failure analysis and not the only one.

Electrical stress data may also provide a design basis for estimating and correlating ESD reliability between tests.

8.2.1.1. Correlation Between EOS and ESD   Since electrical analysis is quicker, another technique used to quickly estimate ESD robustness or devices is applying EOS stress using either transmission lines or easily available laboratory signal generators [Diaz, 1992]. The main advantages of this technique are as follows:

- Robustness is easy to measure and analyze and the devices are less susceptible to parasitics.
- The analysis is more quantitative, allowing model building and development [Beebe, 1996].
- Easier comparison of design specifications is allowed.

Electrostatic discharge stress devices in the 1–2 kV range for 1–100 ns interval whereas EOS stress devices in the microsecond range for tens of volts. Due to this, the damage mechanisms and protection are different. Generally EOS stress shows larger area damage than ESD damage. This larger damage is due to more uniform conduction of current and greater heat spreading in the microsecond time frame in the nanosecond range of an ESD pulse. Usually EOS failures are due to damaged junctions. The relation between the damage threshold and the pulse power–time duration is shown in Figure 8-12.

When the pulse is fast (1–10 ns), as in a CDM pulse, there is not much time for the power to diffuse away from the heat generation area. For example, consider a 10-ns pulse with a thermal diffusivity D of $0.13 \, cm^2/s$; then the distance over which the heat spreads is $\sim 0.36 \, \mu m$ ($\sqrt{Dt}$). An adiabatic approximation is sufficient, relating the power and pulse duration reciprocally, and it is shown in Figure 8-12. When the time is a little larger (~ 1000 ns), the heat has some time to diffuse ($\sim 3.6 \, \mu m$) during the heat generation period. This is the diffusion model proposed by Wunsch and Bell, and here the power is reciprocally related to the square root of the pulse duration [Wunsch, 1968]. Then there is a region of constant power where the heat dissipation and generation are in equilibrium and the device remains operational.

This idea can be extended to measurements. A controlled pulse is applied to the device under test (DUT), and the device is leak tested after the pulsing. Repeated exposure with large pulse width or higher power is applied until the device fails. This can be repeated until a series of data points relating power to the pulse width are obtained. In Figure 8-13 a simplified test setup is shown [Diaz, 1992].

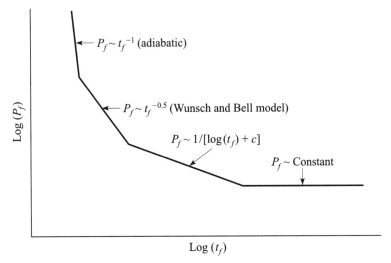

Figure 8-12. Typical regions for power–time failure during ESD/EOS stress.

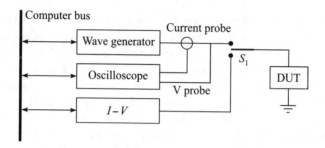

Figure 8-13. Simple EOS measurement setup.

The results of such an experiment are shown in Figure 8-14. Device 1 shows a typical diffusion-type behavior. The EOS region can be extended to the ESD region, and a good estimate of power can be found. Now in the comparison of devices 2 and 3, if one estimated the EOS performance based on ESD performance, they would be the same. However, device 2 shows faster roll-off and is less EOS robust than device 3. One possible reason can be that the power clamps designed for the ESD regime (submicrosecond) shut off or are damaged in the EOS regime (microseconds to milliseconds) [Ramaswamy, 1996], thereby allowing stress buildup.

Generally, EOS stress may be extrapolated to an ESD regime to obtain a reasonable (or pessimistic) answer. This can be seen from the graph in Figure 8-14, where extrapolating from a shallower slope yields lower power than can be tolerated at ESD time frames (~ 1–$150\,\mathrm{ns}$). However, extrapolating from the ESD regime to estimate EOS performance may be erroneous. The EOS measurements should be done separately.

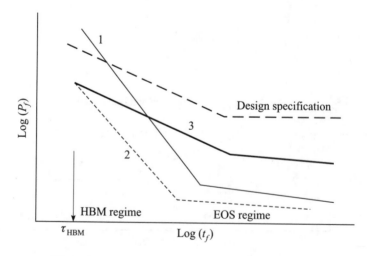

Figure 8-14. Power–time curves for three devices.

8.2.1.2. HBM and MM Correlation Another method to shorten the evaluation time by obtaining quick estimates is to use one ESD test and estimate the robustness of the scheme by a simple correlation formula. This method was explored by Kelly [1995], and results indicate that the MM and HBM show a very close correlation, as shown in Table 8-1. This indicates that if a component passes a certain voltage (V_{mm}) in the machine model mode, then it is likely to pass 11.73 times V_{mm} in the HBM test. Both the machine model and the HBM model have ESD pulse widths that are in the > 10 ns range, thereby allowing heat diffusion to occur.

The correlations between the HBM and CDM and the MM and CDM are poor. A CDM test should be conducted to evaluate the component robustness and HBM or MM results should not be extrapolated. The poor correlation can be attributed to the adiabatic rapid nature of CDM heat generation, and thus little heat diffusion occurs.

8.2.2. Microscopy

A crucial aid in failure analysis is the microscope, which is used to visually identify the damaged site and analyze the damage section (e.g., [Antinone, 1986; Richards, 1992]). A number of techniques are available to the failure analyst, and these are listed in Table 8-2. Of these, the light microscope is the most convenient and a very versatile tool to identify gross defects. The sample preparation is easiest and usually nondestructive. Resolutions of about ~ 1 μm can be detected, but the main handicap with this microscope is the limited depth of focus at high magnification (1000×).

To improve the depth of resolution, techniques such as scanning light microscopy have been constructed. These operate by rastering an optical beam over the surface to be imaged, each point being serially obtained. The total picture is then computer rendered. These microscopes can have enhanced magnification and simultaneously greatly improve the depth of field while retaining the ease of use of an optical microscope.

TABLE 8-1. Correlation between HBM, MM, and CDM Tests

Model to Model	Correlation Coefficient	Regression Analysis
HBM to CDM	0.28	$V_{HBM} = 1.63 \times V_{CDM}$
HBM to MM	0.92	$V_{HBM} = 11.73 \times V_{MM}$
CDM to MM	0.42	$V_{CDM} = 3.37 \times V_{MM}$

Note: The HBM and MM correlate well, but CDM does not correlate well with any other test. For reference the HBM, MM, and CDM rise times are 10–30, 15–30 and 1 ns, respectively.
Source: Kelly [1995].

TABLE 8-2. Microscopy Available to Failure Analyst and Characteristic Properties

Microscopy	Analysis	Samples	Resolution	Reference to ESD Work
Light microscopy	Surface	All	~1 μm	pp. 25, 28
Secondary electron/EDX	Surface morphology	All samples	~5 nm	p. 39
Voltage contrast	Surface potential	Devices	6–20 nm	p. 40
Back-scattered electron	Atomic number, depth, and morphology	—	<1 μm	p. 34
X-ray	Composition	All	0.5–5 μm	p. 37
EBIC	Junction delineation	Semiconductor	0.5–5 μm	p. 44
Surface acoustic wave	—	—	1.5 mm to wavelength	p. 71
Infrared	Junction	Si	~1 μm	p. 57
Scanning light microscopy (SLM)	Surface	All	~1 μm	pp. 63, 65
SLM–optical beam induced current (SLM-OBIC)	Surface	—	~1 μm	p. 63
Emission microscopy (EM)	Surface	Dielectric	0.5 μm	p. 85

Source: Richards [1992].

The other available tool is the scanning electron microscope. Here a sample preparation is required and the sample is usually destroyed. It is primarily a surface analysis technique, but additional modes such as back-scattered electron (BSE) can be used to obtain atomic numbers of an area under inspection. This atomic number information can be used for atomic identification or to enhance the demarcations and to contrast the picture. Here higher magnifications ($10,000\times$) and larger depth of focus (approximately millimeters) are possible, but the downside is that it is a more involved technique requiring some sample preparation.

It should be noted that SEM techniques have been enhanced significantly. Today, SEM techniques can be used not only to inspect visually a sample but also to examine circuit operation. This is made possible by interlinking the computer and the scanning electron microscope. This utilizes the circuit layout data base loaded into the microscope's computer to navigate the probing electron beam on to the active chip. This analysis can provide voltage/time data on a node as well as the physical condition of a node.

A tool that has been extensively utilized is the light emission microscope for detecting very small leakages in oxides. This technique is based on the idea that current passing through a dielectric will emit faint light. This light can be captured and amplified (electronically), and dielectric leakage can be identified. It is useful down to a $\sim 0.5\,\mu m$ resolution and has been extensively used to study dielectric failures [Richards, 1992; Intel, 1991].

There is a continuing struggle for higher resolution and larger depth of focus. With finer geometry and larger silicon die these will continue to be stretched.

8.3. SUMMARY

In this chapter simple ESD stress generation equipment and techniques to stress the device were discussed. Also failure analysis techniques such as electrical I-V and microscopy were elaborated. Simple ESD stress generation tools are as follows:

- A simple HBM stress tool using commonly available components was discussed. The parasitics are not crucial, so some wire lead lengths (3 in. or 10 nH) can be tolerated.
- A simple CDM stress generation tool was shown. This tool can also be constructed using simple components. The parasitics are important and long wire (> 1 in, or 8 nH) should be avoided. Also a high-speed (i.e., 1–5 GHz bandwidth) oscilloscope is required to capture the CDM waveform.
- A convenient form of ESD stress is the transmission line and its construction was shown. This tool is capable of fast edge rates since it has small parasitics. The transmission line data can then be correlated to HBM (and maybe CDM) performance.
- Burn-in testing is essential in VLSI testing today, and it adds additional constraints in ESD design. The cut-in voltage of any clamps must be higher than the highest burn-in voltage. The burn-in environment is also very noisy so clamps have to be made robust enough to minimize any chances of latchup.
- Failure analysis is usually done late in the product cycle and therefore leads to solutions that are usually expensive to correct. However, parts do fail and they need to be screened quickly. Electrical tests are usually the quickest, but usually these are followed by microscopy to validate the electrical diagnosis.

REFERENCES

[Amerasekera, 1991] A. Amerasekara, L. V. Roozendaal, J. Bruins, and F. Kuper, "Characterization and Modeling of Second Break Down for the Extraction of ESD-Related Process and Design Parameters," *IEEE Trans ED*, **ED38**, 1991, p. 2161.

[Amerasekera, 1992] A. Amerasekara and A. Chatterjee, "An Investigation of BiCMOS ESD Protection Circuit Elements and Applications in Submicron Technologies," *Proc. EOS/ESD Symp.*, 1992, p. 265.

[Amerasekera, 1995] Amerasekera and C. Duvvury, *ESD in Silicon Integrated Circuits*, Wiley, West Sussex, England, 1995, p. 18.

[Antinone, 1986] R. J. Antinone et al., "Electrical Overstress Protection for Electronic Devices," Noyes, Park Ridge, NJ, 1986, p. 128.

[Avery, 1991] L. R. Avery, "Beyond MIL HBM Testing How to Evaluate the Real Capability of Protection Structures," *Proc. EOS/ESD*, 1991, p. 120.

[Beebe, 1996], S. G. Beebe, "Methodology for Layout Design and Optimization of ESD Protection Transistors," *Proc. EOS/ESD*, 1996, p. 265.

[Dangelmayer, 1990] G. T. Dangelmayer, *ESD Program Management*, Van Nostrand Reinhold, New York, 1990.

[Diaz, 1992] C. Diaz, C. Duvvury, S. M. Kang, and L. Wagner, "Electrical Overstress (EOS) Power Profiles: A Guideline to Qualify EOS Hardness of Semiconductor Devices," *EOS/ESD Symp.*, 1992, p. 88.

[Diaz, 1994a] C. H. Diaz, S. M. Kang, and C. Duvvury, *Modeling of Electrical Overstress in Integrated Circuits*, Kluwer Academic, Boston, 1994, p. 39.

[Diaz, 1994b] C. H. Diaz, C. Duvvury, and S. M. Kang, "Studies of EOS Susceptibility in $0.6\,\mu m$ NMOS ESD I/O Protection Schemes," *J. Electrost.*, **33**, 1994, p. 273.

[Duvvury, 1988] C. Duvvury, R. N. Rountree, and G. Adams, "Internal Chip ESD Phenomena beyond the Protection Circuit," *IEEE Trans. ED*, **35**(12), 1988, p. 2133.

[Euzent, 1991] B. L. Euzent, T. J. Maloney, and J. C. Donner II, "Reducing Field Failure Rate with Improved EOS/ESD Design," *Proc. EOS/ESD Symp.*, 1991, p. 59.

[Gabriel, 1996] C. T. Gabriel and S. R. Nariani, "Correlation of Antenna Charging and Gate Oxide Reliability," *J. Vac. Sci. Tech. A*, **14**(3), 1996, p. 990.

[Geiser, 1996] G. Geiser and M. Haunschild, "Very Fast Transmission Line Pulsing of Integrated Structures and Charged Device Model," *Proc. EOS/ESD*, 1996, p. 85.

[HBM, 1993] ESD Sensitivity Testing: Human Body Model (HBM)—Component Level, ESD Association Standard, S5.1, 1993.

[Hu, 1996] C. Hu, G. P. Li, E. Worley, and J. White, "Consideration of Low-Frequency Noise in MOSFET's for Analog Performance," *IEEE EDL*, **17**(12), 1996, p. 552.

[Kelly, 1995] M. Kelly, T. Diep, S. Twerefour, G. Servais, D. Lin, and G. Shah, "A Comparison of ElectroStatic Discharge Models and Failure Signatures for CMOS Integrated Circuit Devices," *Proc. EOS/ESD Symp.*, 1995, p. 175.

[Keytek, 1990] Keytek Instrument Corp., "ESD Testing for ICs," Keytek Instrument Corp., Wilmington, MA, 1990.

[Kim, 1992] S. U. Kim, "ESD Induced Gate Oxide Damage During Wafer Fabrication Process," *Proc. EOS/ESD Symp.*, 1992, p. 99.

[Maloney, 1985] T. Maloney and N. Khurana, "Transmission Line Pulsing Technique for Circuit Modelling of ESD Phenomena," *Proc. EOS/ESD Symp.*, Sept, 1985, p. 49.

[Maloney, 1990] T. J. Maloney, "Enhanced P$^+$ Substrate Conductance in the Presence of NPN Snapback," *Proc. EOS/ESD Symp.*, 1990, p. 197.

[Maloney, 1992] T. J. Maloney, "Integrated Circuit Metal in the Charged Device Model: Bootstrap Heating, Melt Damage, and Scaling Laws," *Proc. EOS/ESD Symp.*, 1992, p. 129.

[Maloney, 1993] T. J. Maloney, "Integrated Circuit Metal in the Charged Device Model: Bootstrap Heating, Melt Damage, and Scaling Laws," *J. Electrostatics*, **31**, 1993, p. 313.

[McCarthy, 1990] A. M. McCarthy, W. Lukaszek, L. Larson, and J. McVitte, "Applications of a New Wafer Surface Charge Monitor," *Proc. EOS/ESD Symp.*, 1990, p. 182.

[MMSPEC] ESD Sensitivity Testing; Machine Model (MM)—Component Level, ESD Assoc. Standard S5.2, 1994.

[Oryx] Automated CDM Tester Model 9000, Oryx Instruments and Materials Corp., Fremont, CA.

[Park, 1996] D. Park, C. Hu, S. Zheng, and N. Bui, "A Full-Process Damaged Detection Method Using Small MOSFET and Protection Diode," *IEEE EDL*, **17**(12), 1996, p. 560.

[Pierce, 1988] D. G. Pierce, W. Shiley, B. D. Mulcahy, K. E. Wagner, and M. Wunder, "Electrical Testing of a 256K UVEPROM to Rectangular and Double Exponential Pulses," *Proc. EOS/ESD Symp.*, 1988, p. 137.

[Polgreen, 1992]. T. L. Polgreen and A. Chatterjee, "Improving the ESD Failure Threshold of Silicided N-MOS Output Transistors by Ensuring Uniform Current Flow," *IEEE Trans. Electron Dev.*, **39**(2), 1992, p. 379.

[Ramaswamy, 1995] S. Ramaswamy, P. Raha, E. Rosenbaum, and S. M. Kang, "EOS/ESD Protection Circuit for Deep Submicron SOI Technology," *Proc. EOS/ESD Symp.*, 1995, p. 213.

[Ramaswamy, 1996] S. Ramaswamy, C. Duvvury, A. Amerasekera, V. Reddy, and S. M. Kang, "EOS/ESD Analysis of High-Density Logic Chips," *Proc. EOS/ESD Symp.*, 1996, p. 285.

[Renninger, 1991] R. G. Renninger, "Mechanism of Charged-Device Electrostatic Discharges," *Proc. EOS/ESD Symp.*, 1992, p. 127.

[Richards, 1992] B. P. Richards and P. K. Footner, *The Role of Microscopy in Semiconductor Failure Analysis*, Oxford Microscopy Handbook, Vol. 25, Oxford University Press, 1992.

[Wunsch, 1968] D. C. Wunsch and R. R. Bell, "Determination of Threshold Failure Levels of Semiconductor Diodes and Transistors Due to Pulse Power Voltages," *IEEE Trans. Nucl. Sci.*, **N5–15**(6), 1968, p. 244.

CHAPTER 9

CONCLUSION

The intention of this chapter is twofold: to summarize the key points in the preceding chapters and to remind the reader once again that this book only attempts to systemize ESD, I/O, and related fabrication. There are significant open questions and the second section points to those. It is hoped that this book will provide a first step in answering these questions.

9.1. SUMMARY

It is a classical problem that the book is linearly written and read, and so some points must precede others. In the case of ESD and I/O topics, the discussion on ESD has preceded that on I/O. In reality, ESD protection circuits exist to protect I/O, and they in turn significantly affect I/O design. Both have to be comprehended together. The ESD sections precede I/O sections only out of the necessity of linear writing process and for no other reason as they are heavily interdependent.

In this book a *simplified* and *designable* approach to both I/O circuit design and ESD protection is presented, especially in light of the interdependencies between the fabrication process, I/O requirements, and ESD protection. The book attempts to highlight such interdependencies.

Chapter I provided a historical perspective followed by discussion of the scope of ESD and I/O problems. It was shown that if ESD protection is not methodically approached, the sheer number of IC , their size, and the types of I/O needed are increasingly challenging tasks to undertake in the future.

Understanding these concerns, a simple solution of providing a "current path" was examined in Chapter 2 together with existing examples. Component pieces needed to construct such a current path were elaborated. These components could operate in breakdown or non-breakdown modes. The breakdown components included devices such as thick field oxide clamps, grounded-gate clamps, and silicon-controlled rectifier clamps among others. The non-breakdown-oriented clamps were based on simple diodes, bipolar junction transistors, and MOSFETs. The non-breakdown devices are easier to simulate, analyze, and design, therefore, they have been repeatedly emphasized.

Additional factors affecting ESD and I/O performance such as package and die capacitance were discussed in Chapter 3. These contribute to significant ESD stress reduction and should be continually used in the future. This is attractive, especially since the power requirements of future chips will require significant bypass capacitors both on-chip and on the package. However, small chips will still exist and they will have small capacitances. The ESD threat to small chips will be a major issue in future technologies. The benefit of distributed clamps was also illustrated.

This book recognizes that ESD protection is intricately linked with I/O. Therefore, basic I/O requirements and concepts were presented in Chapter 4. The common clock scheme in which one system clock is used to clock I/O transactions between the chips in a system was described. Their interrelationship with the topology and choice of edge rate with respect to the stub length was discussed. For higher frequency operations or for significant bus lengths the time-of-flight constraint made the common clock system unworkable. The limit imposed by having one system clock for I/O and another approach using a cotransmitted clock was examined. Since the data are sent along with its latching clock, this scheme eliminates the major portion of the common clock limitations. The better the management of skews between the data and the cotransmitted clock, the better the system performance.

Alternative schemes such as pulse width modulation have been examined for situations where the bus has a long settling time but has a far better edge placement accuracy. In this case the pulse width can encode a number of bits that can be recovered and decoded by the receiver. Using this encoding, fewer bus toggles are required but more information is transmitted.

The bus activity generates noise that must be controlled. Schemes utilizing analog and digital compensation methods have been discussed. The idea of compensation is to ensure the I/O buffer performs in a predictable and similar fashion for all types of fabrication processes (fast or slow), voltage variations, and temperature variations. Without compensation the edge rates of I/O buffers can vary by a factor of 2–4 and the output impedance by a factor of 2–3. The compensation can adjust the output buffer impedance and the slew rate to a far lower variation, thereby greatly enhancing bus performance.

With I/O schemes moving toward higher bandwidth rapidly, the requirements placed on ESD are indeed challenging: allowable granularity, low capacitance loading, small size, and voltage compatibility, to name a few.

Once these ESD and I/O requirements are met, the design has to be transformed into layout, and this was discussed in Chapter 5. The device layout is critical for ESD protection, and by far the output NMOS is the most sensitive element. Various device layout options were discussed, including the historical nonsilicided technology and the current silicided versions. For the current silicided technologies the lack of any ballasting action in the N^+ silicide leads to current filamentation in the NMOS source to drain and subsequent melting leading to failure. The NMOS snapback and second breakdown can happen rapidly and may not allow other NMOS fingers in parallel to share the ESD current. Consequently, increasing the NMOS device size rarely works. The need to distribute the current uniformly is therefore critical. Two methods have emerged: to reintroduce the ballast resistor and to lower the snapback voltage below the second breakdown voltage, thus allowing all NMOS legs to snap back and share the ESD current. The ballast resistors can be implemented by using special process masking steps that block out the silicide in the output transistor drain or utilizing the N-well resistor. For snapback voltage modulation the gate-coupled NMOS or the ratioed-gate NMOS scheme can be employed.

Other devices such as the PMOS, waffle transistors, diodes, and SCR layouts have also been discussed, but these are of less critical nature. One key parameter in ESD layout is the uniformity of the layout. This enables better current distribution leading to better ESD performance. The reason to have a 10-μm^2 minimum metal to handle a 1-kV CDM pulse is also discussed. This metal cross-sectional area should be suitably increased if the metal constricts in some regions due to fabrication limitations such as metallization over steps.

The ESD protection and the I/O interact electrically also, and these were discussed in Chapter 6. Broadly these impact minimum output buffer sizing, noise coupling, and I/O signal integrity. The ESD rules can impose a fixed- and rigid-geometry output transistor size, which makes the desired output buffer sizing difficult. It is also seen that if the I/O noise is substantially above the diode cut-in voltages, the core and power supply are not totally isolated. Due to the forward biasing of the diodes, some noise is injected into the power supply. The method to control such coupling involves introducing more diodes into the power supply coupling diode chain. This was a possibility in higher voltage I/O systems such as the 5-V and the 3.3-V generations. With the lower voltage for scaled technologies, the noise is smaller and it is harder to forward bias a diode, and therefore the coupled noise due to ESD coupling diodes is small. The larger decoupling capacitors in the peripheral planes may provide a larger component of coupled noise. Another I/O and ESD interaction occurs between large over- and undershoots of an I/O and the clamping due to the ESD diode. Generally this clamping helps settle the net sooner and also helps reduce the "ringback" voltage. In this clamping process, however, significant current may be dumped into the receiver core power supply, therefore causing noise. Another threat with the forward biased diodes is the possibility of latchup which requires good guard rings around the diodes.

A more recent phenomena examined in Chapter 7 is mixed-voltage I/O which is becoming more prevalent in computer systems. This voltage scaling has been necessary to support finer lithography. In turn, it has led to I/O and core voltages that may be different as chips communicating may be from different process generations. The designs of input and output buffers, pre-drivers, and ESD protection options have to be modified. Further it should be noted that this mixed-voltage interface will become increasingly common and such designs will need to be streamlined.

After the design, layout, and fabrication cycle, the ESD sensitivity of the final circuits has to be measured. For this purpose sophisticated tools exist, but these may be too cumbersome for the laboratory. In Chapter 8, simpler tools like transmission line testers and simplified HBM and CDM apparatuses were described. These tools are easily created, saving time and resources. Also, electrical and microscopic failure analysis were also briefly discussed.

Chapter 9 concluded the book.

9.2. FUTURE QUESTIONS

Significant questions that have only been partially addressed or not addressed at all by this book should be addressed in the future. Some of these are as follows:

- Silicon on insulator (SOI) is expected to replace high-performance circuits in the coming years. Some issues have been raised and initial work has been started in ESD protection. Some concepts illustrated in this book using the current path approach will work; however, due to the very small silicon volume, the diode and the NMOS are less robust than in bulk silicon. So wider devices may have to be implemented. How is the complete SOI ESD protection going to be implemented?
- Controlled collapse chip collapse (C4) will introduce another new packaging dimension to ESD protection of VLSI circuits In wire-bonded packages, the wire bond adds inductance (~ 3 nH) to the ESD path, thereby increasing the path impedance. Now with "cleaner" packages, this impedance is far less, consequently exposing the chip to higher stresses. How does the chip handle the increased stress?
- Multichip module (MCM) protection can be enhanced for internal nodes that are external to the chips but internal to the MCM. What kind of ESD rules or specifications should be applied or developed?
- Capacitive loading due to ESD structures slows down I/O performance. With I/O performance reaching the gigahertz range, very fast input edge

rates are required which makes the capacitance reduction imperative. The Z_0C charging and discharging slow the input signal, which is not very conducive to high-frequency I/O. How can this capacitance be minimized and still meet the ESD requirements?

- Automated design and layout validation will increasingly become important because the growing chip size, metal layers, and number of types of I/O will make manual inspection and verification unreliable. Making smaller blocks and having each ESD safe has been suggested as one method to divide the problem. Even so the final integration has to be verified. How can a method be developed for automated design and layout validation?

- Voltage compatibility from older technologies is already an issue. If the technological generation of the communicating chips is greater than two, then the solutions are difficult. When the difference between the latest and the old technologies is even greater, how can all the timing and ESD requirements be consistently met?

The list is not complete and is only a representation. It is hoped this book provides a first step in the design of future I/O and ESD devices.

Here is a closing thought on a possible direction for I/O and ESD. Considerable effort is needed when simultaneously managing I/O speed, mixed-voltage requirements, the backbiasing issue, area requirements, and ESD reliability. A technology that can simultaneously fulfill all these requirements is optics. High-speed optical I/O designs are in existence today and should be able to address the bandwidth requirements for VLSI. Optoelectronic components are used for isolating systems with high-voltage differentials between two circuits or for isolation and safety. This property can therefore address the mixed-voltage technology requirements. If most of the I/O was done using optics, the need for I/O pads would decrease, thus reducing the ESD hazards as well. The only pins needing protection would be the power pins, which are considerably easier to protect than an I/O pad. A major drawback is that silicon does not readily support optoelectronics. To support optoelectronics directly, a new technique for silicon is required or compound semiconductor interfacing circuits can be used. Therefore, optoelectronics, if achieved efficiently, can be related to a large number of issues raised in this text.

It is hoped that this text has been able to provide the basics of I/O and ESD design. Previously these data were available in fragmented manner in the literature and considerable time and effort on the part of a newcomer were needed to piece it together. We also invite all readers to correct any errors and provide any other helpful suggestions that may enhance the material of this book.

INDEX